: Titles in
ibutions in American History

hic Land Apart: Reassessing Southerners and Their History
avid Smith and Thomas H. Appleton, Jr., editors

orporate City: The American City as a Political Entity, 1800–1850
d P. Curry

cracy Reformed: Richard Spencer Childs and His Fight for Better
nment
d Hirschhorn

Newspaper Opinion on the American Civil War
M. Blackburn

v Jackson and His Tennessee Lieutenants: A Study in Political Culture
n A. Ratner

nism and Radicalism in a Reforming America: 1888–1918
s Robert Shor

enate Munitions Inquiry of the 1930s: Beyond the Merchants of Death
ew Ware Coulter

rontier in the Colonial South: South Carolina Backcountry, 1736–1800
Lloyd Johnson, Jr.

rs of the Spirits: The Judicial Response to Prohibition Enforcement in
, 1885–1935
. Guthrie, Jr.

rt Hoover and Franklin D. Roosevelt: A Documentary History
y Walch and Dwight M. Miller, editors

er Profit and Loss: The British Army and the Fur Traders, 1760–1764
S. Dunn, Jr.

sm and Christianity in Early 20th Century America
H. Dorn, editor

The Final Frontiers, 1

Recen
Cont

The Final Frontiers, 1880–1930

Settling the Southern Bottomlands

John Solomon Otto

Contributions in American History, Number 183
Jon L. Wakelyn, Series Editor

GREENWOOD PRESS
Westport, Connecticut • London

Library of Congress Cataloging-in-Publication Data

Otto, John Solomon.
The final frontiers, 1880–1930 : settling the southern bottomlands / John Solomon Otto.
p. cm.—(Contributions in American history, ISSN 0084-9219 ; no. 183)
Includes bibliographical references and index.
ISBN 0-313-28963-8 (alk. paper)
1. Southern States—History—1865–1951. 2. Southern States—Environmental conditions. 3. Mississippi River Valley—History—1865- 4. Mississippi River Valley—Environmental conditions. 5. Land settlement—Southern States—History. 6. Cotton growing—Southern States—History. 7. Alluvial plains—Mississippi River Valley—History. I. Title. II. Series.
F215.086 1999
976'.04—dc21 99-21710

British Library Cataloguing in Publication Data is available.

Library of Congress Catalog Card Number: 99-21710
ISBN: 0-313-28963-8
ISSN: 0084-9219

First published in 1999

Greenwood Press, 88 Post Road West, Westport, CT 06881
An imprint of Greenwood Publishing Group, Inc.
www.greenwood.com

Printed in the United States of America

The paper used in this book complies with the Permanent Paper Standard issued by the National Information Standards Organization (Z39.48–1984).

10 9 8 7 6 5 4 3 2 1

To the memory of Father John G. Bals,
late rector of St. John's Episcopal Church

Contents

Preface: Settling the Southern Bottomlands

This book examines the settlement history of the alluvial bottomlands of the lower Mississippi River Valley from 1880 to 1930. Although the alluvial bottomlands of northwestern Mississippi, northeastern Louisiana, eastern Arkansas, and southeastern Missouri contained the richest soils in the American South, settlers in the Southern bottomlands faced a host of environmental problems, including seasonal floods, swampy soils, dense forests, physical isolation, insect and arachnid pests, and endemic human and livestock diseases. Given these environmental hindrances, the alluvial bottomlands were still sparsely settled frontiers as late as 1880. Yet, during the next five decades, the swamplands of the lower Mississippi River Valley had been largely transformed into plantations, farms and towns, and the Southern bottomlands had become one of the most densely settled areas in the American South.

Because this is the first history of the Southern bottomlands that focuses on the crucial settlement period from 1880 to 1930, this book should prove useful to historians, social scientists, and general readers who are interested in U.S. environmental, economic, and social history. It is based on primary and secondary sources from the disciplines of history, geography, sociology, agronomy, and ecology. These sources came from libraries and archives in Washington, D.C., Mississippi, Louisiana, Arkansas, Missouri, and elsewhere. Although most of the primary and secondary sources that are cited in the endnotes were written before 1940, I tried to cite as many recent secondary sources as possible.

Drawing on a variety of sources, this book explores the settlement history of the Southern bottomlands—an area that has fascinated me since childhood when I lived only two miles from the Arkansas River bottoms. During the 1950s, my family resided in a large Arkansas county that was a microcosm of the American South. One portion of the county was mountainous terrain that resembled Appalachia, another section was dissected upland that recalled the southeastern piedmont, and the third portion was a tract of Arkansas River bottomlands that was an outlier of the lower Mississippi River Valley.

My family operated a cattle farm in the upland section of the county—a region of red clay that had once been planted in cotton and corn. Decades of row-cropping had depleted and eroded these clay soils. When cotton prices collapsed in the early 1930s, the scarred upland fields could no longer support most of the cotton-farming families, who fled the countryside during the Great Depression. The surviving upland farm families abandoned cotton, and they switched to cattle, hogs, and poultry. By the 1950s, there were no farmers raising cotton in the upland section of the county.

Although the upland farmers in the county had forsaken cotton, the bottomland farmers were still planting cotton on virtually all of the cleared land. Because the alluvial soil was too rich and too level to abandon cotton, this crop retained its grip on the bottomlands. The bottomland towns consisted of little more than cotton gins, general stores, churches, schools, and the occasional honky-tonk. Beyond the bottomland towns lay the flat cotton fields, which were dotted with houses, shacks, barns, and sheds. The cotton fields were the property of the landowning farmers, or "landlords," who worked their acreage with White renters, Black sharecroppers, and Black wage workers. Although renters, croppers, and workers still did the "cotton chopping" (thinning and weeding) as well as the cotton picking, modern machines were part of the cotton economy of the 1950s. Tractors plowed the land and planted the cottonseed. In turn, crop-dusting biplanes buzzed the cotton fields, poisoning the boll weevils and other pests. Finally, huge bulldozers cleared the remaining timberlands in the bottoms, preparing more acreage for cotton farming. I never saw any of the new cotton-picking machines during the 1950s. They were probably too expensive for the local landlords.

Despite the presence of modern machines, bottomland cotton farming in the county was more than one century old. The first cotton farmers entered the bottomlands during the 1840s. Seduced by the fertile alluvial soils, they planned to raise cotton, corn, and livestock in the Arkansas River bottomlands. Hoping to prosper, they soon confronted the realities of life in the Southern bottomlands. The first cotton farmers faced spring floods, sodden soils, thick forests, poor transportation, hordes of pests, and the threat of malarial fevers. After enduring a year or two of such conditions, many farmers fled the bottomlands, and moved to the uplands, where they escaped the floods and other plagues. Yet, some slave-holding farmers remained in the bottomlands. Employing their slave laborers, these farmers built levee embankments to hold back floodwaters, they cleared forest vegetation from the land, and they cut roads through dense forests to reach steamboat landings along the river. Although slaves and farmers suffered from malaria, and although insect pests attacked their cotton fields, these antebellum settlers were adapting to one of the more forbidding natural environments in North America. The Civil War and Emancipation disrupted this settlement process, but many bottomland farmers retained their lands, even though they lost their slaves. These landowning farmers were joined by the "moneyed men" (businessmen), who also acquired bottomland acreage. Because slavery had ended, landlords recruited Black and White tenants and laborers to clear and cultivate their bottomland farms. Hoping to protect their farms from flooding, landlords supported the levee districts that built imposing embankments to hold back floodwaters. To drain the swampy lands, they created drainage districts that dug the ditches for dewatering sodden soils. The construction of local railways and graveled roadways eased the isolation of bottomlands life. Finally, quinine and other medicines offered some relief from malarial fevers, and commercial poisons combated the insect pests, including the troublesome boll weevils that arrived in the early twentieth century. This cotton-farming adaptation was completed by the mid-twentieth century, when landlords employed modern machines to clear lands, prepare seedbeds, and control insect pests.

This settlement process, which unfolded in the bottomlands of the county from the mid-nineteenth to the mid-twentieth centuries, was repeated in similar fashion in dozens of other counties and parishes in the lower Mississippi River Valley. In order to examine this settlement process, I selected a sample of bottomland counties and parishes for statistical comparison between 1880 and 1930—the crucial fifty-year period, when the swamplands were largely transformed into plantations, farms, and towns. Northwestern Mississippi, or the Mississippi Delta, contained eleven counties (Bolivar, Coahoma, Humphreys, Issaquena, Leflore, Quitman, Sharkey, Sunflower, Tallahatchie, Tunica, and Washington), which in 1930 lay within the Mississippi–Yazoo basin. In turn, northeastern Louisiana, or the Louisiana Delta, counted eight parishes (Concordia, East Carroll, Franklin, Madison, Morehouse, Richland, Tensas, and West Carroll), which in 1930 lay within the Ouachita–Tensas basin. In neighboring eastern Arkansas, or the Arkansas Delta, there were fifteen counties (Chicot, Clay, Craighead, Crittenden, Cross, Desha, Greene, Jackson, Lee, Mississippi, Monroe, Phillips, Poinsett, St. Francis, and Woodruff), which in 1930 lay within the Ouachita–Tensas basin, the Arkansas River bottomlands, the White River bottomlands, and the St. Francis basin. Finally, there were six counties (Dunklin, Mississippi, New Madrid, Pemiscot, Scott, and Stoddard) in southeastern Missouri, or the Missouri Delta, which in 1930 lay within the St. Francis basin.[1] Although these 40 counties and parishes in the Southern bottomlands provided the statistical underpinning for this book, I referred frequently in the text to several bottomland communities—Bolivar County, Mississippi; Concordia Parish, Louisiana; Mississippi and Poinsett counties, Arkansas; and Mississippi County, Missouri—which were the subject of detailed histories. The histories of these communities best exemplified the settlement process in the Southern bottomlands during the late nineteenth and early twentieth centuries.

1

Settling the Southern Bottomlands Before 1880

> We travelled the whole of yesterday on this bottom and [we are] not through yet. It is a dead level all the way and densely covered in forest, except a very few small clearing[s] and some newly built log cabins. I suppose it is almost impossible to make any headway clearing; the trees are so very large. There are some very great hinderence [hindrances] to the settlement of this Co. [county], although it is so very rich.[1]

The alluvial bottomlands of the lower Mississippi River Valley presented the most promising but also the most challenging agricultural frontier in the American South. Above the junction of the Red and Mississippi rivers lay extensive bottomlands that covered most of northwestern Mississippi, northeastern Louisiana, eastern Arkansas, and southeastern Missouri. These Mississippi River Valley bottomlands were bounded on the eastern and western margins by bluff escarpments. Since the Mississippi River hugged the escarpment bluffs in Kentucky and Tennessee, there were only small tracts of floodplain in these states. But in northwestern Mississippi, the escarpment bluffs veered eastward, leaving a vast floodplain between the Mississippi and Yazoo rivers. This was the famous Mississippi–Yazoo basin—an alluvial land of streams, bottoms, and ridges that lay between the rivers. The swampy bottomlands of the basin were drained by the Sunflower River, Deer Creek, and Steele's Bayou, which flowed into the Coldwater–Tallahatchie–Yazoo system. In turn, the Yazoo system flowed southward to join the Mississippi River. Below the junction of the Yazoo and Mississippi rivers, the Mississippi River hugged the escarpment bluffs into southern Louisiana. But in northern Louisiana and southern Arkansas, the bluff escarpment veered westward. Between the escarpment and the Mississippi River lay the Ouachita–Tensas basin—an alluvial land of bottoms and ridges that was drained by Bayou Bartholomew and the Ouachita–Tensas–Black system. Flowing southward, Bayou Bartholomew joined the Ouachita River. Receiving waters from Bayou Boeuf, the Ouachita in turn joined the Black River, which was formed by the union of the Tensas River and Bayou Macon. The Black emptied into the Red River and ultimately into the Mississippi River. Reaching into southeastern Arkansas, the Ouachita–Tensas–Black system drained the alluvial lands below the Arkansas River. Above the Arkansas River, the bluff escarpment veered northeastward. Between the escarpment and the Mississippi River lay the prairies, ridgelands, and bottomlands of eastern Arkansas. North of the Arkansas River bottomlands was the Grand Prairie, a treeless upland terrace that formed the western boundary of the White River bottomlands. These swampy bottomlands were drained by the Black, Cache, and White rivers that emptied into the Arkansas River. Beyond the White River bottomlands lay Crowley's Ridge—a timbered ridgeland that intersected northeastern Arkansas. East of Crowley's Ridge was the swampy St. Francis basin, which was drained by the St. Francis and Little rivers. Reaching into southeastern Missouri, the St. Francis system flowed into the Mississippi River. Taken together, the St. Francis basin, the White River bottomlands, the Arkansas River bottomlands, the Ouachita–Tensas basin, and the Mississippi–Yazoo basin comprised the greatest tract of alluvial bottomlands in the American South.[2]

Covering most of southeastern Missouri, eastern Arkansas, northeastern Louisiana, and northwestern Mississippi, the alluvial bottomlands of the lower Mississippi River Valley were subject to seasonal floods. Flooding was possible during the months between December and June, when the tributaries of the Mississippi system dumped floodwaters into the lower Mississippi Valley. Flooded rivers poured out of their banks, inundating bottomlands with silt-laden waters. Muddy waters deposited heavier silt along the riverbanks, creating natural levees of fertile sandy loam that were called *frontlands*, but floodwaters carried the lighter silt onto the bottoms, creating flats of rich clay loam that were called *backlands*. Although the soils of the backlands were richer in nutrients than the frontlands, the poorly drained backlands required draining before farming, whereas the well-drained frontlands needed only clearing before planting.[3]

The rich soils of the Southern bottomlands proved attractive to antebellum settlers, but the dense vegetation, malarial fevers, and physical isolation discouraged settlement. Thick hardwood forests and dense cane-brakes choked the bottomlands. Cottonwood trees crowded the frontlands, gums and swamp oaks covered the seasonally flooded backlands, cypress and tupelo dominated the swampy backlands, and oak, hickory, and dogwood trees covered the ridgelands. In addition, thick stands of giant canes often appeared on the backlands and ridgelands. Because this dense vegetation had to be removed before these fertile soils could be farmed, the bottomland forests posed a formidable barrier to agricultural settlers.[4] These bottomland forests also provided an ideal habitat for anopheline mosquitoes—the insects that transmitted malaria, a debilitating disease that was characterized by chills and fevers.[5] Not only did agricultural settlers have to battle malarial fevers and dense forests, but many also endured physical isolation. During the rainy months, the dirt roads in the bottomlands turned into muddy quagmires, whereas the dry months left many bottomland rivers with too little water for steamboat navigation.[6] Given such environmental hindrances, it was not surprising that settlers did not flock to the alluvial bottomlands during the early nineteenth century.

Extinguishing the Native American land claims in the lower Mississippi River Valley, the federal government acquired millions of acres of alluvial bottomlands during the early nineteenth century. Amassing huge tracts of bottomlands in Mississippi, Louisiana, Arkansas, and Missouri, the federal government hoped to sell this swampland acreage to speculators and settlers. Yet, antebellum investors often looked askance at the Southern bottomlands, which had a well-deserved reputation for floods and fevers. Consequently, swampland acreage found relatively few buyers before 1849. Holding a surplus of public swampland, the federal government donated unsold acreage to the states in the lower Mississippi River Valley. The Swampland Acts of 1849 and 1850 transferred some 24 million acres of federal swamplands to the state governments of Mississippi, Louisiana, Arkansas, and Missouri. Under the terms of these acts, states were free to sell swamplands in order to finance flood control and land drainage schemes.[7]

Receiving a land bonanza, the states of Mississippi, Louisiana, and Arkansas planned to sell swamplands to speculators and settlers. The state of Missouri, however, transferred its swampland acreage to the counties by 1857, and these local governments sold lands to investors. Offering swamplands for as little as $1.25 an acre, state and county governments sold thousands of acres of bottomlands during the flush years of the 1850s.[8]

The Swampland Acts of 1849 and 1850 accelerated the settlement of the lower Mississippi River Valley. As an example, Bolivar County, Mississippi, counted only 2,577 inhabitants in 1850, but the county held 10,471 residents by 1860. Between 1850 and 1860, Bolivar's slave population grew from 2,180 to 9,078, the tilled land increased from 16,973 to 85,188 acres, and cotton production rose from 4,723 to 33,452 bales. Thanks to increased production and population, land values in Bolivar County soared from $739,000 in 1850 to $8,759,000 by 1860.[9]

During the 1850s, Bolivar and other bottomland counties and parishes attracted thousands of Southern agriculturalists, including slave-owning planters, slave-holding farmers, and slaveless farmers. Owning more than twenty slaves, "planters" possessed the laborers needed to develop bottomland acreage. Purchasing bottomlands from governments or from speculators, slave-owning planters usually acquired estates of more than five hundred acres. By 1860, plantations that measured more than five hundred acres were common in northwestern Mississippi, northeastern Louisiana, and southeastern Arkansas. Such plantations, however, were rare in northeastern Arkansas and southeastern Missouri. These areas were the realm of slave-owning farmers, who held fewer than twenty slaves, and slaveless farmers, who depended on their families for labor.[10]

This distribution of farmers and planters in the Southern bottomlands was reflected in the population statistics from the 1860 federal census. In 1860, census takers counted 31,625 residents in the Missouri Delta counties, but only 13 percent of these people were slaves. Because Missouri was a "border slave state," the institution of slavery was less secure than in the deep South states. Additionally, southeastern Missouri lay along the northern border of the "cotton belt," so it had barely enough frost-free days to grow cotton. Given the short growing season and the insecure status of slavery, the Missouri Delta attracted few cotton planters before 1860. Most cotton planters sought bottomlands in the Arkansas, Louisiana, and Mississippi deltas, where the growing season was longer and where slavery was freer from abolitionist influences. Thanks to the influx of cotton planters, slaves were far more numerous in the Arkansas, Louisiana, and Mississippi deltas than in the Missouri Delta. By 1860, slaves were 43 percent of the 76,737 residents in the Arkansas Delta counties, composed 81 percent of the 78,587 inhabitants in the Louisiana Delta parishes, and were 84 percent of the 57,862 residents in the Mississippi Delta counties.[11]

Moving into the Mississippi, Louisiana, and Arkansas deltas with their slaves, planters found cheap fertile land, which offered higher cotton yields than other soils in the Old South. The fresh soils of Tensas Parish, Louisiana, for example, yielded over one bale of cotton per acre without any manuring. Although the backlands offered the highest cotton yields, these poorly drained, swampy areas required costly drainage as well as clearing before cultivation. In contrast, the better drained frontlands and ridgelands offered lower yields, but these higher locations required only clearing before cultivation. Because the frontlands and ridgelands required far less "improvement" than the backlands, these higher, better drained areas proved most attractive to antebellum settlers.[12] As an example, when Frank Montgomery, a planter from the Natchez area, settled in Bolivar County, Mississippi, in 1855, he found an "unknown wilderness," which had only "a few plantations on Lake Bolivar and Egypt Ridge" as well as "a few plantations along the banks of the [Mississippi] river."[13]

Although the frontlands and ridgelands offered better drained soils, these higher locations were choked with trees, underbrush, and canes. To clear this dense vegetation, planters employed their axe-wielding slaves.[14] A former slave from Mississippi recalled the land-clearing practices of the time. During the late summer, slaves went into the forests to " 'deaden' or girdle the trees on a given tract." Using axes, they cut "through the bark all around the trunk about thirty inches from the ground." The "trees so treated soon died and in a year or two were in condition to be removed." The "season selected for clearing the land was winter, beginning with January." The "trees, except the larger ones, were cut down, cut into lengths convenient for handling and piled into great heaps, called 'log heaps,' and burned." The "undergrowth was grubbed out" with hoes and then "piled and burned." The "chopping was done by the men slaves and the grubbing by women." Clearing cane-brakes proved easier than removing brush and trees. The "cane was so thick that when they were clearing new ground, it would have to be set on fire, and the crackling that would ensue was like the continuous explosion of small fire crackers." After "the new land was cleared, corn was put in," and the "next season" cotton was planted. Cotton "was seldom planted on newly cleared land, as the roots and stumps rendered it difficult to cultivate the land without injury to the growing plant."[15]

In addition to clearing "new land," slaves also labored on the levee embankments that lined the river banks. These homemade earthen embankments held back the spring floods in most years.[16] When he arrived in Bolivar County in 1855, Montgomery noted that the plantations were all partly protected by small private levees, since "the entire country was annually inundated by floods which came down the river every spring."[17]

Building levee embankments and clearing forest lands, slave laborers literally carved plantations from the Southern bottomlands. These newly created plantations were "consolidated" estates, where the slaves worked in groups, or gangs, under the close supervision of overseers. At night, the slaves lived in villages, or quarters, near the overseers' and planters' dwellings. Living in quarters and working in gangs, slave laborers harvested bountiful crops of cotton and corn from newly opened fields.[18]

Among these new bottomland plantations was Amanda Stone's Brokenburn estate in Madison Parish, Louisiana. Brokenburn was "a newly opened place," when Stone purchased the tract. Because the new estate had a sawmill as well as a cotton gin, her slaves quickly constructed dwellings and other plantation buildings. The Stone family lived in a large, eight-room house with long halls and galleries. Across the yard, there were "two long rows of [slave] cabins, facing each other across a broad sweep of Bermuda grass." At the end of the cabin rows was "the overseer's four-room log house," which was "rough but substantial." Near the overseer's house stood the cookhouse, stables, barns, and other outbuildings. Working 150 slaves on her thirteen hundred-acre estate, Stone was among the wealthier planters in the Southern bottomlands.[19]

Although the Stones resided on their Brokenburn estate, many wealthy planters lived outside the Southern bottomlands, and they left the daily management of their estates to overseers. This was certainly the case in Concordia Parish, Louisiana, where many of the planters resided in the

Natchez area of Mississippi. By 1860, there were so many absentee planters in Concordia Parish that a traveler claimed that "the actual white population" was mostly "a population of overseers."[20] Yet, even in Concordia Parish, the bottomlands contained some slave-owning and slaveless farmers. In Concordia, the wealthy planters claimed the frontlands along the Mississippi River, but small farmers resided on the frontlands along the Black River, where they raised crops in leveeside fields and grazed livestock in the swamplands.[21]

Slave-owning and slaveless farmers lived throughout the Southern bottomlands, but they were most numerous in northeastern Arkansas and southeastern Missouri.[22] Among such bottomland farmers was Ed Elder. Losing his Mississippi farm because of indebtedness, Elder moved to eastern Arkansas, where he settled in what later became Cross County. Arriving in Arkansas, Elder took up "160 acres of land under the [state] law granting titles to swamp lands to actual settlers, who had to do a certain amount of ditching." Although he owned no slaves, Elder "put up a good comfortable building, planted an orchard, and gathered around him a few horses, cows, hogs and sheep." Yet, the hogs and sheep had "to be penned up at night to prevent them from being feasted on by the wolves."[23]

Bottomland farmers not only had to battle predators, but they also had to endure malarial chills and fevers. At this time, antebellum Southerners believed that malaria was caused by "bad air," or "miasmas," which were released by decaying vegetation in the swamplands. They regarded the swampland mosquitoes, which were the actual carriers of malaria, as pests and "tormentors." Suffering from malaria, bottomland residents downed concoctions made from dogwood bark, cherry bark, and whiskey.[24] Such homemade remedies failed to ease the malarial fevers and chills. When Henry Morton Stanley, who later achieved fame as an African explorer, clerked in Altschul's store at Cypress Grove, Arkansas, in 1860, he claimed that few customers "did not bear some sign of the pernicious disease [malaria] which afflicted old and young in the bottom-lands of Arkansas." After clerking for a week, Stanley also succumbed to malaria. He dosed himself with quinine—an expensive but effective treatment for malaria.[25]

In addition to malaria, bottomland residents also had to cope with the major floods that periodically struck the Southern bottomlands. Major floods swept down the Mississippi River Valley in the years of 1844, 1847, 1849, 1850, 1851, 1858, and 1859. Breaching private levees, these floods inundated fields, swept away improvements, and drowned livestock.[26] The 1850 flood was particularly disastrous for low-lying Concordia Parish, Louisiana, which was bordered by the Mississippi, Red, Black, and Tensas rivers. Following this spring flood, the Police Jury passed an ordinance to enclose Concordia Parish with levees. Paying local planters to do the construction work, the parish completed "thousands of yards of levee" by the spring of 1851. During the spring flood of that year, there was "a feeling of safety from an overflow more than has been enjoyed [in] any previous year."[27] In addition to locally built levees, the states of Louisiana, Arkansas, and Mississippi used public swamplands, which they had received from the federal government, to fund some levee construction. Levee contractors, who hired Irish laborers to do the shovel work, received swampland scrip as payment. In turn, southeastern Missouri counties, which had received swamplands from the state, offered swampland scrip to levee builders. Thanks to state, local, and private efforts, an estimated two thousand miles of levees, which averaged eight to ten feet in height, lined both banks of the Mississippi River by 1858. The great flood of 1858, however, breached these low levees, flooding much of the bottomlands.[28] Following this devastating flood, the state of Mississippi created the "1858 Levee District" to coordinate levee building in the Mississippi Delta. Funding construction with levee taxes and swampland scrip, the district began rebuilding the levee line in the Mississippi Delta. By 1861, the district had built dozens of miles of embankments that offered stronger barriers to future floods.[29]

Braving major floods, many bottomland residents also struggled with poor transportation. Although planters living along the Mississippi River had access to reliable river transportation, the settlers who lived along the tributaries depended on seasonal river transportation to ship cotton to market. These isolated planters and farmers clamored for railways, which provided reliable, year-round transportation.[30] Responding to these calls, several companies attempted to build railroads into the Southern bottomlands. The Memphis & Little Rock railway began construction in eastern Arkansas, the Cairo & Fulton began building in southeastern Missouri, and the Vicksburg, Shreveport & Texas railroad began work in northeastern Louisiana. Yet, no antebellum railroad companies entered the forbidding Mississippi–Yazoo basin in northwestern Mississippi.[31]

Despite poor transportation, periodic floods, and malarial fevers, agricultural settlement of the Southern bottomlands quickened during the 1850s. The counties and parishes in northwestern Mississippi, northeastern Louisiana, and southeastern Arkansas, which attracted many slave-owning planters, grew at a faster pace than the counties of northeastern Arkansas and southeastern Missouri, which lured fewer planters.[32] The settlement of the Southern bottomlands, however, ground to a halt in the spring of 1861, when the United States plunged into the Civil War (1861–1865).

During the Civil War, Union navies and armies invaded the lower Mississippi River Valley, planning to bisect the Confederacy. Bottomland planters, who resided in the riverside counties and parishes, found themselves living in a battleground. Contending Union and Confederate forces pillaged plantations and destroyed levees.[33] Fleeing the destruction, many planters abandoned their riverside homes, leaving behind "princely estates" and "hundreds of Negroes" [*sic*].[34] Although some bottomland slaves fled with the refugee planters, more sought safety behind Union lines, becoming "contrabands" who served the federal cause. Occupying the Mississippi River Valley by 1863, Union forces acquired thousands of willing contrabands as well as hundreds of abandoned plantations. Hoping to grow cotton for the Northern textile mills, federal authorities leased abandoned plantations to Northern newcomers who hired contrabands to raise crops.[35] Yet, this wartime attempt to resurrect the bottomland cotton industry achieved meager results. Leased plantations became targets for Confederate guerrillas, who harassed Northern lessees, seized Black workers, and burned cotton gins. Many of the lessees in the lower Mississippi Valley had to abandon their estates, leaving unharvested cotton in the fields. Although leased plantations near Union garrisons escaped guerrilla attacks, these estates often fell prey to caterpillars, or "Confederate worms," which devastated cotton plants.[36] Thanks to cotton worms and Confederate guerrillas, many bottomland plantations lay untilled. By war's end, much of the bottomland region was a desolate landscape of abandoned plantations, overgrown fields, and ruined levees. Because the levee system was in disrepair, spring floods swept across the bottomlands.[37] Crossing the flooded Mississippi Delta in the spring of 1865, a party of Texans, who were returning home after the war was over, found that the "great spread of improved land was lying idly vacant" and that only "small portions" of several plantations "had been planted."[38]

Returning to his Mississippi home when the Civil War ended in May 1865, Confederate cavalryman Frank Montgomery crossed the flooded Delta in a boat, reaching his "Beulah" plantation in Bolivar County. He found his plantation house was in ruins, his wife was living in a log cabin within the ruins, and only a dozen slaves were still residing on his estate. His wife, nevertheless, planned to raise a cotton crop, and she promised the remaining slaves one fourth of any cotton that they raised. Arising the next morning, Montgomery "assembled all the negroes [*sic*], and told them the southern armies had all surrendered, the war was over, and they were free." He also promised to honor "the contract they had made with their mistress," and Montgomery "advised them to stay" and raise the cotton crop.[39]

Although returning bottomland planters tried to raise cotton crops in 1865, they faced formidable problems. Because many abandoned plantations were in the hands of the Bureau of Refugees, Freedmen and Abandoned Lands, which administered abandoned plantations for the federal government, the returning planters had to obtain presidential pardons in order to reclaim their estates. Although most plantations were eventually reclaimed by prewar owners, the estates were in ruins. The work stock was missing, tools were gone, slave laborers were now free, and there was no working capital.[40] Planters needed capital to repair estates, hire free laborers, and resume cotton planting. Facing such obstacles, some planters despaired for the future. They offered to sell their estates for a fraction of prewar prices.[41] Land agents advertised dozens of prime bottomland estates, hoping to attract Northern buyers, who wished to pursue cotton planting. As an example, an agent in Jackson, Mississippi, advertised the following estate in the Louisiana Delta: "Plantation on Tensas River, La.; Madison parish [*sic*], 1000 [*sic*] acres, 800 cleared. Sold for $60 [in] gold [per acre] before the war. Price is now $30 [in] greenbacks per acre."[42] A more popular option than selling land was leasing acreage to Northern newcomers who wanted to plant cotton. Because cotton prices soared to astronomical heights by 1865, many Northerners wished to lease Southern cotton plantations, make a fortune growing cotton, and then return to the North. By leasing estates to Northern newcomers for one to three years, bottomland planters hoped to earn the capital they needed to resume cotton planting, when the Northern lessees returned home.[43] The most popular option, however, was to borrow capital from Northern financiers or Southern merchants and then resume planting as soon as possible. Borrowing funds at high interest rates, bottomland planters rebuilt their estates, hired workers, and tried to raise cotton.[44]

Whether they were Southern- or Northern-born, planters and lessees faced the same hazards after the war. Spring floods, cotton worms, and labor troubles doomed the 1866 and 1867 cotton crops. Following these disastrous crop seasons, most Northern newcomers returned home, and the surviving Southern planters plunged deeper into debt, as cotton prices fell in 1867. Some relief arrived in 1868, when the federal government repealed the 2.5-cent per pound excise tax on cotton and cotton prices rebounded.[45] Good cotton seasons from 1868 to 1872 enabled many planters to avoid financial ruin. Despite a measure of prosperity, planters struggled to pay the high land taxes that Republican state governments imposed during the Reconstruction years (1868–1877). Although planters forfeited much swampland to the states for nonpayment of land taxes, Democratic state governments, which supplanted the Republicans by 1877, slashed land taxes.[46] Gaining relief from high land taxes, planters endured another round of hard times following the Panic of 1873. Cotton prices plunged to new lows, driving some planters into bankruptcy. Creditors foreclosed on estates, indebted planters sold lands, and numerous plantations changed hands.[47] Despite financial travails, some bottomland planters retained their estates, and some even added land to their plantations. These surviving planters were joined by financiers, who had foreclosed on bankrupt plantations, and by newcomers, who purchased devalued bottomland estates.[48]

Whether they were newcomers or survivors, bottomland planters worked their estates with a mixed labor force of wage hands, sharecroppers, and renters.[49] Some planters preferred to work their estates with wage hands who labored for monthly or yearly wages. Operating by the "wages system," planters controlled their workers, supervised the farm work, and kept their estates "in good repair." Despite these advantages, planters found it difficult to hire enough wage hands because Black laborers preferred sharecropping and land renting to wage working. Under the "share system," planters furnished "the land, quarters, wood, privileges, farming implements, stock and feed," while the croppers provided the labor for raising the crops. After the harvest, planters and croppers shared the harvest "on the basis agreed upon at the beginning of the year," which was "in most cases one-half." Finally, under the "rental system," planters rented acreage "for a specified amount of money, or pounds of cotton." Renters and croppers assumed more risk than wage workers, but they worked at their own pace, laboring fewer hours in the fields than wage hands, and they earned more income than hirelings when crops were bountiful.[50]

Although the bulk of renters, croppers, and hirelings on bottomland plantations were Blacks, some poor Whites also labored on swampland estates. Bottomland planters, who were desperate for agricultural labor, offered higher wages to laborers and more generous terms to croppers and renters. Hoping to earn more money, poor Blacks and Whites moved into the Southern bottomlands after the war. Working on swampland plantations, Black and White wage hands, croppers, and renters formed the labor class that rebuilt bottomland agriculture during the postbellum years.[51]

The shift from slave laborers to free laborers after 1865 left its mark on the bottomland plantations. Before the war, slaves lived in quarters and worked in gangs. But after the war, freed people were reluctant to live and work like slaves. In order to retain Black workers, planters had to make concessions by allowing many of them to crop or rent tracts of land.[52] Planters frequently subdivided their estates "into small tracts—commonly called 'parcels' or 'cuts'—each of which was operated by a tenant," and they built dwellings on these tracts to house the tenant families. Thanks to this subdivision, the old consolidated plantations of slavery times evolved into "tenant plantations," where the land was divided into numerous tracts that were worked by croppers and renters. Subdividing much of their land into tenant tracts, planters often retained "a considerable quantity of improved land," which they worked with wage hands.[53] Because plantations were now divided into "home farms," which planters worked with wage hands, and into "tenant tracts," which planters operated with tenants, the "fragmented" plantations of the postbellum years scarcely resembled the consolidated plantations of antebellum times.[54]

Working their fragmented estates with free laborers, bottomland planters weathered labor shortages and economic depressions after the war, but they suffered from the floods that struck the Mississippi River Valley. In 1865, 1867, 1874 and 1882, major floods swept down the valley.[55] Because the Civil War had damaged many of the levees along the bottomland rivers, Southern states faced the difficult task of rebuilding levees to keep out floodwaters.[56] In Missouri, the county governments failed to rebuild the levees after the disastrous flood of 1858, and there was little levee construction until the 1880s.[57] But in Arkansas, the state government shouldered the burden of levee reconstruction. In 1869, the state legislature assigned levee construction to the Commission on Public Works and Internal Improvements. If the majority of landowners in an area wanted levees, the state commission was to conduct surveys, make estimates, and award contracts for the levee work. Unfortunately, the state commission awarded many contracts to fraudulent firms, and there was little levee construction. The Arkansas bottomlands remained vulnerable to major floods.[58] A similar scenario occurred in Louisiana, where the state government also assumed responsibility for levee construction. After the costly flood of 1867, levee repair became the province of the Board of Public Works in 1868. Because the board accomplished little work, the state government approved the sale of levee bonds in 1870, and this prompted some levee building. But in 1871, the state government turned to the Louisiana Levee Company. This private corporation planned to rebuild the state's levees, using public land taxes to fund the work. Although the Levee Company absorbed millions in tax monies, little levee work was done. In 1877, the state legislature assigned the task of levee building to the Board of State Engineers. The board wasted less tax money than the Levee Company, but it accomplished even less work. Despite extravagant expenditures, state efforts to repair Louisiana's levees failed miserably. Parish governments managed to rebuild some of the levee line, but Louisiana's levees were still in disrepair by the 1882 flood.[59] In contrast to Louisiana and Arkansas, where state governments failed to reconstruct the levees, Mississippians made a more successful attempt to restore the levee system. In November 1865, planters in the southern counties of the Mississippi Delta formed a levee board to repair the riverside levees. The board imposed a local land tax, it also taxed the local cotton crops, and the board issued bonds to fund levee work. Private contractors employed Irish shovelers, who did the repair work during the winter months. Recognizing the board's work in 1877, the state legislature created the Mississippi Levee District, which included the southern Delta counties. Taxing the local cotton crop, the district funded the repair of old levees and even built new embankments.[60] The northern Delta counties, however, relied on private levees to keep out floodwaters. When private levees proved inadequate, the state legislature chartered Levee District No. 1 in 1871, which included the northern Delta counties. Surviving for only five years, the demise of this levee district left the northern Delta vulnerable to major floods.[61] Facing the threat of flooding, concerned Mississippians called for federal aid to restore the Delta's levee system. The U.S. Congress finally responded in 1879, creating the Mississippi River Commission. Although the commission was empowered to improve navigation on the Mississippi River, it also aided levee boards in repairing levee embankments. Thanks to commission aid, Delta levees averaged seven feet in height by 1882. But in that year, a great flood swept down the Mississippi River Valley, severely testing these rebuilt levees.[62]

Enduring major floods every few years, many bottomland planters also faced transportation problems. Although planters who lived along the Mississippi River had dependable steamboat transportation to the Memphis, New Orleans, and St. Louis markets, those planters who lived along the Mississippi's tributaries had less reliable transportation to these regional markets.[63] On major tributaries such as the Arkansas, Ouachita, and Yazoo rivers, small steamboats ascended to Little Rock, Arkansas; Monroe, Louisiana; and Greenwood, Mississippi during much of the year.[64] Yet, on such lesser tributaries as the St. Francis, Tensas, and Sunflower rivers, steamers entered only during the "high water" months. Planters living along these lesser streams had to store cotton until waters were passable, or they had to haul cotton bales over swampy roads to riverports on navigable streams.[65]

Railways offered more reliable transportation than waterways, but few bottomland planters had access to railroads. Before the Civil War, northeastern Louisiana, eastern Arkansas, and southeastern Missouri held three partially completed railroads. The Vicksburg, Shreveport & Texas (V,S&T) railroad spanned the bottomlands of northeastern Louisiana, running from Vicksburg on the Mississippi River to Monroe on the Ouachita River. In turn, the Memphis & Little Rock (M&LR) railway ran from Memphis on the Mississippi River to Madison on the St. Francis River. By 1862, the M&LR completed another section between Little Rock on the Arkansas River and Devalls Bluff on the White River. The swampy gap between Devalls Bluff and Madison was serviced by steamboats that operated between these riverports. Finally, the Cairo & Fulton (C&F) railroad ran from Bird's Point on the Mississippi River to Sikeston in the Missouri Delta. The Civil War, however, left these three bottomland railways in shambles.[66] When the war ended, a syndicate of Monroe, Louisiana businessmen purchased the defunct V,S&T, which they renamed the North Louisiana & Texas (NL&T) railroad. Despite frequent floods, the NL&T railway restored the line from Vicksburg to Monroe by 1870. Suffering financial reverses, the company was unable to build the final link between Monroe and Shreveport. The troubled NL&T passed into the hands of U.S. capitalists, who renamed it the Vicksburg, Shreveport & Pacific (V,S&P) railroad. Continuing to lose money, the V,S&P became the property of English capitalists by 1881. Despite its financial troubles, the V,S&P provided a serviceable rail route across the Louisiana Delta during the 1870s.[67] In contrast to the V,S&P, the M&LR railway enjoyed greater financial success after the war. Receiving state aid, the M&LR rebuilt the old line and then closed the gap between the White and St. Francis rivers by 1871. Although seasonal floods often shut down parts of the railroad, the M&LR provided an overland route from Little Rock to Memphis, and market towns sprang up along this swampland artery.[68] In addition to the M&LR, the Arkansas bottomlands were also serviced by two postbellum railways: the Arkansas Central, a narrow-gauge railroad that ran from Helena on the Mississippi River to Clarendon on the

White River, and the Little Rock, Mississippi River & Texas (LR, MR&T) railroad, which stretched from Chicot Point on the Mississippi River to Pine Bluff on the Arkansas River. Begun in 1871, the Arkansas Central struggled financially. After foreclosure, it was reorganized as the Arkansas Midland in 1878. The LR,MR&T proved more financially successful than the Arkansas Central, but it was subject to frequent flooding. The company moved the track to a higher roadbed that ran from Pine Bluff to Arkapolis (Arkansas City) on the Mississippi River. Completing the link from Pine Bluff to Little Rock by 1881, the LR,MR&T provided another route across the Arkansas Delta.[69] Finally, the partial C&F railroad, which had deteriorated during the war, reverted to the state of Missouri in 1866. The next year, the state sold the C&F line to a firm that represented the St. Louis & Iron Mountain (St.L&IM) railway. Dubbing it the Cairo, Arkansas & Texas (C,A&T), the new owners extended the old C&F line to Poplar Bluff, Missouri, where it intersected a branch of the St.L&IM railway. In 1874, the C,A&T was formally consolidated with the St.L&IM to form the St. Louis, Iron Mountain & Southern railway. In addition to the St. L,IM&S, the Missouri Delta also contained the Little River & Arkansas Valley (LR&AV)—a narrow-gauge railway that ran from New Madrid to Malden.[70] Although the Missouri, Arkansas, and Louisiana deltas possessed several railways by 1880, the Mississippi Delta had only two short narrow-gauge lines—the Mobile & Northwestern (M&NW) and the Greenville, Columbus & Birmingham (G,C&B). The M&NW crawled from Dowd's Landing to Clarksdale, and the G,C&B ran from Greenville to Johnsonville. Totaling only seventy miles in length, these two railways served only a tiny fraction of the Mississippi Delta by 1880.[71] Because railroad service was still inadequate in the Southern bottomlands, many swampland planters continued to rely on seasonal riverways and swampy roadways to transport cotton bales to market.

Thanks to poor transportation, frequent flooding, and other problems, the Southern bottomlands were still sparsely settled as late as 1880. In that year, the U.S. Census Office counted 390,014 inhabitants in the counties and parishes that lay within the Mississippi, Louisiana, Arkansas, and Missouri deltas. Covering 24,180 square miles of territory, the bottomland counties and parishes contained only sixteen people per square mile (see Table 1). Because the population density was less than eighteen people per square mile, the Southern bottomlands were still "extensive agricultural frontiers," which characterized an "early stage of settlement" by planters and farmers.[72] Despite possessing the richest soils in the American South, the Southern bottomlands remained agricultural frontiers, whose development would require five decades, millions of dollars, and thousands of settlers.

2

The Southern Bottomlands During the Late Nineteenth Century (1880–1900)

> Before the War the real estate of this parish was assessed at something over seven millions of dollars, and the personal [property], including slaves, for about seven and one half millions, and now the real property is assessed at one million, and the personal [property] at about three hundred thousand dollars. The reason of this vast depreciation is caused by the change in the system of labor, and [by] the dilapidated condition of the levees, for this parish like all other alluvial districts is dependent upon the levees. With a good levee system, there is no doubt but what the lands in this parish would appreciate millions in a very short time.[1]

In 1880, the U.S. Census Office completed the most thorough census in U.S. history. Publishing numerous volumes of census data in the early 1880s, the Census Office offered the most detailed statistical portrait to date of the U.S. economy and society. Included in this copious flow of census publications were two heavy volumes that were devoted to U.S. agriculture. The first volume covered agricultural production in the United States, but the second volume, which was published in two parts, contained statistical data about the U.S. cotton industry. The first part of this cotton volume, which dealt with the cotton industry in the lower Mississippi River Valley, presented a detailed survey of the cotton-producing counties and parishes in the Southern bottomlands.[2]

In Mississippi, the leading cotton-producing state in the Union, the ten counties (Bolivar, Coahoma, Issaquena, Leflore, Quitman, Sharkey, Sunflower, Tallahatchie, Tunica, and Washington) that lay within the Delta in

1880 (see Table 1), harvested 197,375 cotton bales. The Delta not only produced over one fourth of Mississippi's cotton crop, but also contained the two most productive counties in the state. Washington County, which harvested 54,873 bales of cotton, led all Mississippi counties in cotton production, and Issaquena County, which harvested 0.88 bales per acre, led all the state's counties in cotton yields per acre. Despite this impressive showing, the Mississippi Delta achieved only a fraction of its potential cotton production. In 1880, the delta counties tilled 255,638 acres of cotton land or only 7 percent of the region's total acreage. If the delta had been properly protected from flooding, and if the delta's land had been properly drained, the Census Office claimed that the region could have harvested an estimated crop of 2,250,000 bales. Yet, such wild estimates for future cotton crops were only fantasies in 1880. At that time, the Mississippi Delta had only 109,598 residents, and the Delta was a thinly settled frontier with eighteen people per square mile[3] (see Tables 1, 5, and 7).

The neighboring Louisiana Delta, which counted only 90,686 inhabitants and which had a population density of barely nineteen people per square mile, also planted only a fraction of the potential cotton acreage. The 1880 census revealed that the eight bottomland parishes (Concordia, East Carroll, Franklin, Madison, Morehouse, Richland, Tensas, and West Carroll) harvested 184,116 cotton bales. Producing more than one third of the state's cotton crop, the Louisiana Delta contained the two leading cotton parishes in the state. Tensas Parish, which harvested 41,859 cotton bales, led Louisiana's parishes in cotton production, and East Carroll Parish, which harvested 0.95 bales of cotton per acre, led the state's parishes in cotton yields per acre. In fact, East Carroll had the highest yield per acre of any county in the American South. Although the Louisiana Delta's soils yielded nearly a bale of cotton per acre, the agriculturalists of the Louisiana Delta tilled 223,348 acres of cotton land or only 7 percent of the delta's total acreage[4] (see Tables 1, 5, and 7).

The neighboring Arkansas Delta, which counted only 136,845 residents and which had fewer than fourteen people per square mile, also tilled only a fraction of its potential cotton acreage. The fifteen Delta counties (Chicot, Clay, Craighead, Crittenden, Cross, Desha, Greene, Jackson, Lee, Mississippi, Monroe, Phillips, Poinsett, St. Francis, and Woodruff) produced 183,079 cotton bales or nearly one third of Arkansas' total crop. The Delta also contained Chicot County, which harvested 0.94 bales per acre—the highest cotton yield of any county in the state. The delta counties, nevertheless, tilled only 263,569 acres of cotton land or a mere 4 percent of the total acreage in the Arkansas Delta[5] (see Tables 1, 5, and 7).

In contrast to the Arkansas Delta, which harvested a substantial cotton crop, the six bottomland counties (Dunklin, Mississippi, New Madrid, Pemiscot, Scott, and Stoddard), which lay within the Missouri Delta, produced only 15,357 cotton bales. Yet, this modest cotton crop represented three-fourths of Missouri's total cotton harvest. Because the cotton plant needed at least two hundred frost-free days to mature, and because southeastern Missouri had a growing season of barely two hundred days, the Missouri Delta lay on the northern fringe of the Southern "cotton belt." Despite the threat of untimely frosts, Missourians managed to harvest two thirds of a bale per acre on the fertile delta soils. Although it was possible to raise cotton in southeastern Missouri, Delta farmers planted only 23,448 acres of cotton land or 1 percent of the region's total acreage. In contrast to this tiny cotton acreage, Delta farmers tilled 165,086 acres of corn land, which yielded 5,275,619 bushels of corn. The Missouri Delta's huge corn crop far exceeded the corn harvests of the Mississippi, Louisiana, and Arkansas deltas. Although the Missouri Delta proved ideal for corn cropping, the region counted only 52,885 residents, and the Delta remained a thinly settled frontier with fewer than sixteen people per square mile[6] (see Tables 1, 5, and 7).

Because the Missouri, Arkansas, Louisiana, and Mississippi deltas were still sparsely settled frontiers, only a fraction of the total land in the Southern bottomlands was in the hands of argiculturalists. By 1880, only 30 percent of the total acreage in the Mississippi, Louisiana, Arkansas, and Missouri deltas was farmland, and the remaining bottomland acreage was in public or corporate hands. Not only had agriculturalists claimed a mere fraction of the bottomlands, but they had "improved" little of their farmland. By 1880, only one tenth of the acreage in the Southern bottomlands was "tilled"—a category that included cropland, fallow acreage, and plowable pasture[7] (see Table 5).

Fencing in their tilled lands with split-rail fences, bottomland agriculturalists allowed their "scrub" livestock to forage on the unfenced acreage, or "open range," in the Southern bottomlands. These scrub cattle and hogs, which were descended from European prebreed stock, were small, scrawny animals that survived on native forage during the year. Cattle foraged on native grasses during the warm months and fed on evergreen canes during the brief winters. Hogs, in turn, subsisted on "mast" (acorns, nuts, and roots) during the year. Identifying their range stock with brands and ear marks, bottomland agriculturalists collected and marketed some of these scrub cattle, and they trapped, penned, and fattened some scrub hogs for butchering. Although scrub animals were rather poor in quality, they cost little to raise. Scrubs foraged on the unfenced range during the year, requiring neither fodder nor protection but furnishing meat for sale and consumption.[8]

Ranging scrub livestock on the unfenced range and tilling only a small fraction of the acreage in the Southern bottomlands, agriculturalists congregated on the frontlands, the natural levees that lined the major streams, and on the occasional ridgelands, which intersected the bottomlands. In the Mississippi Delta, for example, the "larger proportion of the population and of the lands under cultivation within the region" were "found along the highlands [frontlands] that border the Mississippi River and along Deer Creek." Other planters occupied "Dogwood Ridge"—the low sandy ridge that lay between the Yazoo and Sunflower rivers. Extending from Coahoma into Leflore counties, the Ridge was "well settled, especially in its northern part." Because agriculturalists lived along Dogwood Ridge, Deer Creek, and the Mississippi River, the rest of the delta was barely settled. The "scanty population and the low percentages of tilled lands in the interior counties such as Quitman and Sunflower" showed "that large portions of the bottom" were "almost without settlements." Not only were the interior counties subject to "the great danger of inundation," but these remote areas also lacked access to reliable transportation.[9]

Transportation played a crucial role in determining settlement in the Southern bottomlands. In the Mississippi Delta, the planters who lived along the Mississippi River in Tunica, Coahoma, Bolivar, Washington, and Issaquena counties enjoyed an advantage in transporting cotton to market. It cost Bolivar County planters only one dollar to send a cotton bale by steamboat down the Mississippi to the New Orleans market. In contrast, planters who lived along seasonally-navigable Deer Creek in isolated Sharkey County, shipped cotton only during the "high water" months from December to June, and it cost as much as $2.75 to send a cotton bale to New Orleans. Agriculturalists living in the interior counties of Leflore, Tallahatchie, Quitman, and Sunflower experienced comparable problems in transporting cotton to market. Given the difficulties in shipping cotton, most agriculturalists in the Louisiana Delta also lived along navigable streams. Planters in East Carroll, Madison, Tensas, and Concordia parishes congregated on the front lands along the Mississippi River, and they shipped cotton to New Orleans for one dollar per bale. Beyond these alluvial parishes lay the interior communities of West Carroll, Morehouse, Richland, and Franklin. Because these interior parishes contained the Bayou Macon ridgelands and the Bastrop hills, they were home to many farmers as well as planters. In these back country parishes, the shipment of cotton to market was either by steamboat on the interior streams to New Orleans or "by rail to Vicksburg and thence by river." It cost little more than one dollar to ship a bale when high water allowed steamers to ascend the interior streams, but it cost as much as $2.50 to send a cotton bale by rail to market. And when the waters were "too low for navigation" on the interior streams, the "railroad to Vicksburg" charged even higher rates. Transportation troubles also prompted many agriculturalists to locate along navigable streams in the vast Arkansas Delta. In Chicot, Desha, Phillips, Lee, Crittenden, and Mississippi counties, planters and farmers occupied the frontlands along the Mississippi River and its tributaries. Enjoying access to the Mississippi River and its tributaries, agriculturalists in these counties shipped cotton by steamboat to the New Orleans and St. Louis markets for less than one dollar per bale. Beyond these riverain counties lay Crowley's Ridge—an extensive ridgeland that covered portions of Clay, Greene, Craighead, Poinsett, Cross, St. Francis, Jackson, Woodruff, and Monroe counties. Lying between the St. Francis and the White River valleys, Crowley's Ridge offered farmers a dry and healthy refuge from the swampy, malarial bottomlands. Despite these advantages, the silty-loam soils of the Ridge yielded only half a bale of cotton per acre, and ridgeland agriculturalists struggled to ship their cotton bales to market. Farmers hauled cotton bales over rutted roads to distant steamboat landings or railway stations. Reaching these shipping points, farmers paid one dollar to ship a cotton bale by steamer to Memphis or $1.65 to send a bale by railway to the Memphis market. Farmers living in isolated Craighead County paid over twice as much to ship a cotton bale by steamboat to St. Louis, Missouri. The agriculturalists who lived in the Missouri Delta also sent much of their cotton to the St. Louis market. In remote Dunklin County, farmers occupied the ridgelands because the bottomlands of the St. Francis and Little rivers were "too low to be available for cultivation without a general system of drainage." Raising cotton on ridges and prairies, Dunklin's farmers hauled cotton bales to distant river landings and rail stations for shipment to St. Louis. It cost as much as $2.25 to haul a cotton bale and another to ship it to market. In neighboring Stoddard County, which lay in the center of the Missouri Delta, farmers hauled cotton bales to the railway for shipment to St. Louis at the cost of $2.50 per bale. But in the riverside counties of Pemiscot, New Madrid, Mississippi, and Scott counties, many farmers occupied the frontlands along the Mississippi River. They shipped cotton bales on steamboats to the St. Louis, Memphis, or New Orleans markets for less than one dollar per bale.[10]

In addition to facing transportation problems, bottomland agriculturalists also had to deal with a shortage of agricultural laborers. In antebellum times, planters and farmers imported slave laborers to work on their bottomland estates. But in the postbellum years, agriculturalists had to recruit free laborers to work on their bottomland plantations and farms. Although some postbellum workers were hill country Whites who drifted into the bottomlands after the war, the majority of agricultural workers were Blacks who emigrated from the southeastern states. Some bottomland planters traveled to the southeastern states to recruit laborers for their estates, but most relied on labor agents to supply them with Black workers.[11] Thanks to the influx of Black laborers after the war, the Black population of the Southern bottomlands grew from 148,691 in 1860 to 236,995 by 1880. The 1880 census revealed that 60 percent of the population was Black.[12] Three fourths of the inhabitants of the Mississippi and Louisiana deltas were Black, and half of the people in the Arkansas Delta were Black, but only one tenth of the residents of the Missouri Delta were Black (see Table 3).

Living in the Southern bottomlands, most Black residents rented lands, cropped on shares, or worked as wage hands on plantations and farms.[13] By

1880, more than half of the farms in the Southern bottomlands were operated by renters and croppers. "Tenants" (renters and croppers) operated 72 percent of the farms in the Louisiana Delta, and tenants worked half of the farms in the Mississippi and Arkansas deltas, but tenants operated only 43 percent of the farms in the Missouri Delta. The remaining farms in the Southern bottomlands were owner-operated, and many of the farm owners relied on wage hands to work their estates (see Table 9).

Despite the influx of Black workers in postbellum times, labor shortages remained a problem in the Southern bottomlands. The census reports on cotton production often mentioned that farm acreage lay idle because of the lack of agricultural labor. In Morehouse Parish, Louisiana, for example, a correspondent from Bastrop reported that "about 10 per cent of [the] land once cultivated now lies out for want of laborers." And in Desha County, Arkansas, a resident of Napoleon claimed that half of the best lands lay "idle for want of labor and not from poverty of soil." And finally, in Pemiscot County, Missouri, a correspondent from Braggadocio noted that none of the frontlands lay "turned out, except for want of labor."[14]

Not only was labor in limited supply, but there was also a shortage of investment capital. Thanks to political turmoil and economic depression, Northern and foreign capitalists were reluctant to invest in the Southern bottomlands before 1880. During the Reconstruction era (1865–1877), the political unrest in the Southern states discouraged many investors. But after 1877, the political uncertainty of the Reconstruction years gave way to conservative Democratic rule. Democratic state governments lowered property taxes, making land investments more attractive. And by 1879, the economic depression, which had followed the Panic of 1873, was virtually over. Economic recovery, in turn, signaled a rise in commodity prices.[15] Higher timber and cotton prices made Southern bottomland acreage a more lucrative investment. The marketable timber taken from bottomland acreage was said to pay for the rich soil that yielded bountiful crops of cotton.[16]

Investors soon discovered that millions of acres of bottomland were available for purchase in the lower Mississippi Valley. This land was in the hands of state governments in Mississippi, Louisiana, and Arkansas, county governments in Missouri, and railroad companies in Louisiana, Arkansas, and Missouri. The state governments of Mississippi, Louisiana, and Arkansas still held some unclaimed swamplands, which the federal government had donated. Missouri counties, in turn, had unclaimed acreage, which they had received from the state. In addition to these public swamplands, the states had acquired private bottomlands, which were forfeited for nonpayment of taxes. Because tax-forfeited lands and unclaimed swamplands earned no tax revenues, state and county governments were willing to sell them for a pittance.[17] Finally, federal and state governments had donated swampland acreage to corporations that were building bottomland railroads. As an example, the V, S&P railway held sizable land grants in northeastern Louisiana; the

M&LR railroad retained grant lands in eastern Arkansas; and thc St.L, IM&S railroad acquired the land grants that had been awarded to the C&F and the LR,MR&T lines in southeastern Missouri and eastern Arkansas. Holding thousands of acres of swamplands, railway companies were anxious to sell land to timbermen and settlers, who would pursue timber cutting and cotton planting along their lines. Railroad corporations would not only earn revenue from land sales, but they would also earn money from freight traffic along their routes.[18]

Among the corporate and public owners of swampland, the state of Mississippi was most desperate to dispose of bottomland acreage. Holding half of the total acreage in the Mississippi Delta by 1877, the state government readily sold tax-forfeited bottomlands for a fraction of the back taxes. Between 1877 and 1881, the state sold 600,000 acres in piecemeal lots. And in 1881, Mississippi disposed of two huge tracts of Delta land. In that year, the state sold 706,000 acres to the Georgia Pacific, a regional railroad that planned to build across the width of the Delta, and 774,000 acres to the Memphis & Vicksburg (M&V), a local railroad that planned to build across the length of the delta. The Georgia Pacific railway, however, soon transferred its Delta acreage to an Anglo-American land syndicate, which was headed by Edward Evers. When Evers' syndicate failed to pay the taxes on its lands, the state reacquired the 706,000-acre tract, offering the land again at public auction. A group of Chicago land investors, who were led by Thomas Watson and sons, acquired 450,000 acres of Evers' land in 1886. Creating the Delta & PineLand Company, the Watsons sold large tracts of bottomlands to timber companies. Not only did the Delta acreage, which the state sold to the Georgia Pacific, quickly pass into the hands of outside investors, but the Delta land that the state sold to the M&V soon became the property of Northern capitalists.[19]

In 1882, a Northern railroad syndicate that was headed by Collis Huntington purchased the M&V railroad as well as its huge land holdings. Acquiring the M&V in Mississippi, Huntington's syndicate began building the Louisville, New Orleans & Texas (L,NO&T) railway between Memphis and New Orleans. Completed in 1884, the L,NO&T provided a year-round, all-weather route across the length of the Mississippi Delta. Yet, the new L,NO&T line soon had competitors. In 1881, the Georgia Pacific railroad acquired the G,C&B—a short, narrow-gauge Delta railway (see chapter 1), and the Georgia Pacific completed the line between Greenville, Mississippi, and Birmingham, Alabama, by 1889. Another competitor was the Illinois Central railroad, which planned a route from Cairo, Illinois, to New Orleans, Louisiana. Acquiring the New Orleans, St. Louis & Chicago (NO,St.L&C) railroad, the Illinois Central projected a branch line from Jackson, Mississippi, to the Delta. Chartered as the Yazoo & Mississippi Valley (Y&MV) railroad in 1882, this branch line skirted the eastern Delta, touching the Yazoo River at Yazoo City and at Greenwood.[20] Emerging in the 1880s, the Y&MV, the Georgia Pacific, and the L,NO&T railways proved crucial in the settlement of the Mississippi Delta. Offering reliable, year-round service, railroads were soon moving Delta lumber and cotton to distant markets.

Railroad construction not only boomed in the Mississippi Delta during the 1880s, but the Louisiana, Arkansas, and Missouri deltas also added hundreds of miles of new trackage. The V,S&P railway finally completed the link from Monroe to Shreveport in 1884, providing a dependable outlet for the timber and cotton of the Louisiana Delta. In addition, the construction of such new railroads as the New Orleans & Northwestern; the Natchez, Red River & Texas (N,RR&T); the Louisiana, Arkansas & Missouri (L, A&M); and the New Orleans, Natchez & Ft. Scott (NO,N&Ft.S) also served the interests of Louisiana's timbermen and farmers. In neighboring Arkansas, Jay Gould, the Northern railroad magnate, acquired the St.L, IM&S railway in 1880, and he hiked freight rates on this important route. Unwilling to pay Gould's rates, James Paramore led a group of St. Louis businessmen who planned to build a narrow-gauge railway from Missouri to Texas. Acquiring the LR&AV railroad (see chapter 1) in southeastern Missouri, Paramore founded the Texas & St. Louis (T&St.L) railroad, which completed the line from Missouri to Texas in 1884. When the T&St.L experienced financial troubles, Samuel Fordyce acquired the narrow-gauge railway, and he switched to standard gauge in 1886. Passing into the hands of Jay Gould in 1888, the line became the St. Louis–Southwestern railway or the "Cotton Belt Route." In addition to the St. Louis–Southwestern, such new railways as the Kansas City, Ft. Scott & Memphis (KC,Ft.S&M), the White & Black River Valley (W&BRV), the Batesville & Brinkley (B&B), the Cache Valley, and the St. Louis, Kennett & Southeastern (St.L, K&SE) served the timbermen and farmers of northeastern Arkansas and southeastern Missouri.[21]

Thanks to new railroad construction in the Southern bottomlands, many isolated areas were opened to timbering and farming. As soon as railroads were constructed, "small towns began to spring up from five to ten miles apart to meet the requirements of surrounding country settlements."[22] These nascent railroad towns invariably included a "few wee shops" with "a draggled stock of haberdashery and groceries," a buzzing sawmill that transformed logs into "weather boarding," and a "railroad platform" with piles of cotton bales.[23]

Typical of these new railroad towns was Cleveland, Mississippi—a delta hamlet on the L,NO&T railroad. Named for President Grover Cleveland, the town had only a half dozen stores and saloons when young William Dockery arrived in 1888. Opening a store in Cleveland, Dockery soon acquired a sawmill and some timberland. Succeeding in the timber industry, he then turned to cotton planting. Dockery amassed nearly ten thousand acres in the Sunflower River Valley, where he established a cotton plantation in 1895. Depending on the seasonal Sunflower River for transportation, Dockery soon built a branch line railroad, dubbed the "Pea Vine," to link his plantation headquarters to the main line railroad at Boyle. On Dockery's plantation, the headquarters included a store, a sawmill, and a cotton gin—three buildings that symbolized Dockery's rise from storekeeper to timberman to cotton planter.[24]

Although Dockery achieved greater success than most newcomers, others sought economic opportunity in the Southern bottomlands. They included Southerners and Northerners, Whites and Blacks, capitalists and paupers. Arriving in the alluvial bottomlands, newcomers sought timber, land, and wages in the areas that were opened by new railways.[25]

Railway corporations, which hoped to earn money from land sales and from freight revenues, encouraged timbermen and agriculturalists to settle along their lines. Railroad companies sold tracts of woodland to timbermen, who in turn set up sawmills along the railways. Sending the prime hardwood logs to the sawmills, timbermen left "cut-over" woodlands that were cluttered with tops, branches, stumps, and uncut trees. Because timber companies had to pay taxes on this cut-over land, they usually disposed of this acreage by selling it to agricultural settlers.[26] Many settlers, nevertheless, did not wait for the timber companies to place cut-over land on the market. Buying railroad land on time, they tried to hack farms from bottomland forests. When Robert and Mollie Beevers arrived at Cleveland in eastern Bolivar County in 1886, they came to a "wild country," which was covered with "heavy timber" and "thick cane." Beneath this thick cane, which stood "fifteen to twenty feet high," the "land was rich as cream." Mollie Beevers claimed that a "good worker could cut one-fourth of an acre a day in the tall cane brakes." Cutting and burning cane proved much easier than removing the heavy timber.[27] To clear swampland forests, settlers girdled the bark to "deaden" the standing trees. When the standing trees in deadenings decayed, settlers felled, piled and burned the dead trees, leaving the rotting stumps in the fields. Plowing around the stumps, they cultivated cotton and corn crops in the partially cleared fields.[28] Because the task of land clearing was so daunting, some settlers abandoned their farmsteads. Although the railroad companies sold land for only "five dollars an acre on long time terms," it was "expensive to clear and many people, both white and black, would quit after a year or two and sell for a small equity." Dockery remembered "seeing one forty acres traded for a cow and another forty acres traded for a Winchester rifle." Yet, some of the bottomland settlers persevered and prospered. They cleared the forest land, harvested cotton crops, and hauled cotton bales over rutted roads to steamboat landings and railway stations.[29]

Although some newcomers sought timberland and farmland, others hoped to find work as timber cutters, sawmill operatives, and agricultural laborers. Transient White and Black men drifted into the bottomlands to cut timber and work in the sawmills. These workers generally moved on to the next job, when they cut-out the timber in an area.[30] In turn, agricultural laborers not only included single men but also families who were seeking economic betterment in the fertile bottomlands. Black Southerners "themselves spread the word of crops 'high as a man on horseback' and of plants so thick with cotton that a first-rate picker could bag five hundred pounds a day."[31]

Labor agents toured the Southern states, recruiting Black workers for bottomland plantations. Seduced by a honey-tongued agent, who portrayed the Arkansas swamps as a land of "wondrous things," the Pickens family of Anderson County, South Carolina, signed a labor contract in January 1888, obligating them to work for an Arkansas Delta planter. Because the Pickens family had no money for travel expenses, the labor agent bought the train tickets for the long journey that ended at a dreary railway station in eastern Arkansas. Clutching "sacks of clothes, quilts, baggage, bundles and baskets," the Pickens family climbed into a freight wagon, which hauled them across a frozen swampland to their new home. Arriving at the plantation during the dead of winter, the Pickenses had no funds to buy food, so they purchased provisions on credit from the planter. Owing the planter for the purchased provisions as well as for the train tickets, the family was already in debt, when they started sharecropping in the spring. Although they raised 50 bales of cotton, the Pickenses failed to erase their debts, and the planter claimed they owed more money. Falling deeper into debt and suffering from "malarial chills and fevers," the Pickens family slaved for another year, making "a bigger and better crop." But "at reckoning time," it was the same story. There was "still enough debt to continue the slavery." Facing a grim future in the Arkansas swamplands, the Pickens family planned an escape. Finding an honest planter near Little Rock, they agreed to rent "a small farm" on his estate. Slipping away from their old plantation at night, the Pickenses escaped "a state of debt-slavery" in the Arkansas bottomlands.[32]

Despite the threat of debt-slavery and malarial fevers, Black families poured into the Southern bottomlands during the 1880s. In one year, a single railroad "brought into the Mississippi Valley over 10,000 Negro [*sic*] immigrants" from "the Atlantic and Gulf States."[33] Fleeing poverty and violence in the upland South, they hoped to find a better life in the bottomlands. Arriving with few possessions, Black newcomers generally worked for wages, or they cropped on shares.[34] If working for wages, field hands received monthly wages, housing, food, and fuel. Wage workers enjoyed fairly steady salaries, but they had little chance for advancement. Share cropping was riskier than wage working, but it offered a better chance for economic improvement. Employers provided sharecroppers with land, housing, work stock, and tools for the crop year, and they advanced supplies to croppers on credit. After the harvest, employers claimed half of the cotton that sharecroppers raised, they deducted the cost of supplies and interest on the debt from the croppers' half of the cotton, and they turned over the remaining revenue to the sharecroppers. If cotton crops were bountiful, and if employers were honest, share croppers cleared some money at the end of the year. By investing this money in "teams, gears [tools] and implements necessary for making a crop," croppers were ready to rent land. Furnishing their own teams and tools, renters leased lands, paying one third or one fourth of the crop (or the equivalent in cash) as rent.[35] Ambitious renters often leased the "deadened lands," where "trees had been killed by girdling and the underbrush burned off." Renting this unimproved land for a pittance, tenants became "temporary pioneers," clearing new fields, harvesting bountiful crops, and turning tidy profits. Clearing more money than share croppers, some renters took the next step, buying land of their own.[36]

Although Black landowners were scattered across the Southern bottomlands, they were particularly numerous along the L,NO&T railway in the Mississippi Delta. The L,NO&T sold railroad lands on generous terms to Black farmers, who settled "at Shaw, Shelby, Duncan, Clarksdale, Lula, Tunica, and Hollywood." The largest concentration of Black landowners, however, was at Mound Bayou in Bolivar County. In 1887, the land commissioner of the L, NO&T railroad encouraged Isaiah Montgomery to create a Black farm colony in the forest around the Mound Bayou railway station. The Black colonists "set bravely to work," building "rude cabins out of split logs," making railroad ties to earn some money, and clearing forests to create fields. Growing rapidly, the colony counted 250 townspeople and 1,500 rural residents by 1891, and Black farmers owned over 15,000 acres around the town of Mound Bayou.[37]

Despite the presence of Black-owned farms in the Southern bottomlands, most large estates, or plantations, were White-owned. Many of the White plantation owners claimed huge estates, having "from 500 to 2,500 acres in cultivation." Although some White planters resided on their bottomland estates, many were absentee owners. Absentees lived in other states, and they employed White managers to supervise daily operations. Some absentee owners even dispensed with managers, and they rented "directly to the negroes [*sic*]."[38]

This was the case on Major R. W. Millsaps' plantation in the Mississippi Delta. Residing in Jackson, Mississippi, Millsaps owned a five hundred-acre estate near Greenville, which he worked with twenty tenant families. Until 1890, "he had a white overseer and he was constantly in trouble of one kind or another with his tenants." Disgusted with this arrangement, Millsaps dismissed his manager, and he rented the land directly to the tenants "on terms that would give both landlord and tenant a reasonable profit." Millsaps never lost a cent by renting land, and his Black renters prospered. Every renter "on the place owned his own mules and waggons [*sic*]" and was "out of debt." With their savings, Millsaps' renters bought houses in the nearby town of Leland.[39]

Worked by renters, croppers, and hirelings, plantations dominated the economic and social life of the Southern bottomlands. An Arkansan, who resided in the Black River Valley during the 1880s, succinctly described the plantations of her era: "A plantation, to-day [*sic*], is generally an estate, a group of little [tenant] farms, rented on shares. A fourth of the cotton or a third of the corn is the usual rental."[40]

The tenants on bottomland plantations were not only cultivators of the soil, but they were also clearers of the forest. An Arkansan, who resided on a Monroe County plantation during the 1880s, recalled the "annual log-rolling," which was held "in the late summer, after the crops were laid by." Forming communal work groups, the Black tenants went into the deadenings to fell the trees and "roll" the logs. After cutting the best timber for fence rails and for firewood, they felled, rolled, and burned the remaining trees "to clear the land for cultivation." Thanks to annual log-rollings, tenants steadily enlarged the cultivated acreage on bottomland plantations.[41]

As plantations spread across the Southern bottomlands, the region became increasingly vulnerable to floods. Flooding now resulted in enormous property losses. Major floods inundated the Mississippi Valley in 1882, 1884, 1890, 1893, and 1897.[42] Given the threat of costly floods, levee districts were created to build and maintain riverside embankments in the Southern bottomlands.

The Mississippi Levee District (MLD), which was created in 1877, served as the model for other bottomland levee districts. Selling bonds and imposing taxes to fund levee work, the MLD contracted to rebuild the levees in the southern delta counties. Contractors hired Irishmen to wield the shovels and wheelbarrows used in levee construction. Although these hand-built levees resisted floodwaters in normal years, they failed during the great flood of 1882. Following this destructive flood, the Mississippi River Commission, which had been created to improve river navigation, began providing federal funds for levee construction in the Mississippi River Valley. Receiving federal aid, the MLD rebuilt the levees in the southern Delta counties, using animal-drawn scrapers to create stronger embankments. These rebuilt levees held back floodwaters during the 1884 inundation. The successful levee-building of the MLD inspired the state of Mississippi to create the Yazoo–Mississippi Delta Levee District (Y–MDLD) in 1884. Responsible for protecting the northern Delta counties, the Y–MDLD began building a coherent levee line in the northern Delta.[43]

The successful levee districts of the Mississippi Delta inspired other states to create districts for flood control. New levee districts in Arkansas and Louisiana not only sold bonds and imposed taxes to fund levee construction, but these districts also received unsold swamplands from the state governments. Selling swamplands to timber companies and land speculators, levee districts raised additional funds for construction. The 1890 flood, nevertheless, broke many of these new levees, and the Mississippi River Commission offered additional aid to the levee districts. Drawing on federal as well as local funds, districts began rebuilding the levee line. By 1897, levees enclosed most of the Mississippi–Yazoo basin, Ouachita–Tensas basin, White

River bottomlands, and St. Francis basin in the Mississippi Valley. The MLD and Y–MDLD protected northwestern Mississippi; the Fifth Louisiana and the Tensas Basin (created in 1886) levee districts guarded northeastern Louisiana; the Chicot (created in 1883), the Clay and Greene (created in 1887), the Laconia (created in 1891), the Red Fork (created in 1891), and the St. Francis (created in 1891) levee districts protected eastern Arkansas; and the St. Francois (created in 1893) levee district guarded southeastern Missouri.[44] The Yazoo, Tensas, and White levees stood more than twelve feet high, but the St. Francis levees were only nine feet high, and there were huge gaps in the St. Francis levee line. In 1897, when the greatest flood on record struck the Mississippi River Valley, there were twenty-three breaks along the incomplete St. Francis line, fourteen breaks occurred in the White River levees, the Yazoo levees suffered six breaks, and four breaks perforated the Tensas line. Following the disastrous 1897 flood, which inundated one third of the Southern bottomlands, local levee districts, which again received aid from the Mississippi River Commission, began rebuilding the "patchwork" levee line. Levee builders now cleared the land before constructing levees, carefully removing all stumps, roots, and logs. After clearing land, they excavated "muck ditches," which were filled with well-tamped clay. These muck ditches not only prevented seepage under levees, but they also anchored the huge levee embankments. Using mule-drawn scrapers, Black workers built embankments that were more than eighteen feet high.[45] Although major floods still threatened the Mississippi River Valley, the rebuilt levees offered better protection for bottomland plantations, farms, towns, and railroads.

Bottomland railways, in particular, benefited from flood control efforts. Railroad construction and consolidation boomed during the late nineteenth century. Fragmentary railways evolved into lengthy lines that in turn became portions of national rail systems. Railroads carried bottomland cotton and lumber to Memphis, New Orleans, and St. Louis for shipment to domestic and foreign markets.

The railroad boom was most pronounced in the Mississippi Delta, where railway trackage soared from seventy miles in 1880 to more than seven hundred miles by 1900. Confronting dense forests and swampy lands, construction crews penetrated Delta swamplands to clear trees, build embankments, and fashion trestles.[46] Despite high construction costs, Delta railways proved profitable, thanks to the lumber and cotton traffic. In fact, the LNO&T railway was so successful that it attracted the attention of the mighty Illinois Central railroad. Planning a route from Chicago to New Orleans, the Illinois Central coveted the L,NO&T, which handled most of Delta's trade. In 1892, the Illinois Central purchased the LNO&T and added it to its subsidiary Yazoo & Mississippi Valley (Y&MV) line. Controlling the north–south routes across the delta, the expanded Y&MV line dominated but failed to monopolize the Delta's trade. The Georgia Pacific line, which became part of the Southern Railway system in 1894, provided the Delta with a competing east–west route that terminated at the Atlantic coastal ports.[47] Thanks to a flurry of railroad construction and consolidation during the late nineteenth century, much of the Mississippi Delta enjoyed efficient rail service by 1900.

In contrast to the Mississippi Delta, railroad building lagged in the Louisiana Delta. Yet, northeastern Louisiana possessed two major railways, which were integrated into national systems. The V,S&P line crossed the St.L,IM&S road at Monroe. The V,S&P was part of the "Queen & Crescent Route" that linked northeastern Louisiana to the New Orleans market, and the St. L,IM&S, which had absorbed such local lines as the New Orleans & Northwestern and the Natchez, Red River & Texas, was part of the Missouri Pacific system that linked northeastern Louisiana to the St. Louis market.[48]

The St.L,IM&S division of the Missouri Pacific also competed with the St. Louis–Southwestern railway (Cotton Belt Route) for the cotton and lumber traffic of eastern Arkansas. The main lines and branches of these roads blanketed eastern Arkansas, claiming most of the region's trade for the St.Louis market. The old M&LR railroad, however, offered Arkansas access to the Memphis market. Although the M&LR was devastated by the 1897 flood, this line became part of the Choctaw, Oklahoma & Gulf (C,O&G) system in 1899. Dubbed the Choctaw & Memphis (C&M), the old Little Rock to Memphis route continued to serve eastern Arkansas. Finally, the Kansas City, Ft. Scott & Memphis railway, which was leased to the St. Louis and San Francisco (Frisco) system in 1901, crossed northeastern Arkansas, offering another route to the Memphis market.[49]

Although eastern Arkansas was well served by major railways, southeastern Missouri still offered a field for such local railroad entrepreneurs as Louis Houck. Between 1880 and 1902, Houck acquired or built the Cape Girardeau, the St. Louis, Kennett & Southern (St.L,K&S), the Kennett & Osceola (K&O), the St. Francois Valley, the Pemiscot Southern, the Missouri & Arkansas (M&A), the Morley & Morehouse (M&M), and the St. Louis, Morehouse & Southern (St.L,M&S) railways. Although Houck sold his railroad "empire" to the St. Louis and San Francisco system in 1902, his local railways had opened much of the Missouri Delta to timbering and farming, and Houck's railroads had augmented the St.L,IM&S and St. Louis–Southwestern systems, which linked southeastern Missouri to the St. Louis market.[50]

Shipping commodities to the St. Louis, Memphis, and New Orleans markets, bottomland planters and timbermen prospered until the hard times of the 1890s. During that decade, overproduction, foreign competition, and financial panics depressed commodity prices to postbellum lows.[51] Cotton, for example, averaged fourteen cents per pound in the 1870s and eleven cents per pound in the 1880s but only seven cents per pound in the 1890s. During the worst years between 1897 and 1898, cotton sold for little more than a nickel per pound. Because it cost as much as a dime to raise one pound of cotton in the hill country, where growers had to apply commercial fertilizers to exhausted fields, cotton was no longer a profitable crop in the upland South. Upland cotton growers either plunged into debt, or they sought bottomlands to grow cotton at lower cost.[52] Lower prices "drove the cotton growing out of the hill country and into the lowlands along rivers," where the bottomland soils did not require expensive fertilizers. Raising cotton for as little as a nickel per pound on "new and unexhausted [*sic*]" bottomlands, most bottomland planters managed to survive the grim years of the 1890s. And when cotton prices rebounded to nine cents a pound in 1900, bottomland growers earned sizable profits. Because it cost only twenty to twenty-five dollars to produce a five hundred-pound cotton bale in the Mississippi Delta planters earned twenty to twenty-five dollars in profits per bale by 1900. Consequently, bottomland growers profited at a time, when many Southern cotton growers struggled.[53] Because swampland soils yielded bountiful cotton crops without the use of costly fertilizers, the Southern bottomlands continued to attract land seekers and job searchers during the 1890s.

Black job searchers and land seekers from the upland South found the bottomlands particularly attractive. Farm laborers in the upland South barely earned fifty cents a day, and sharecroppers harvested only one third of a cotton bale per acre from upland fields. Yet, in the Southern bottomlands, planters paid seventy-five cents for a day's work, and some share croppers harvested up to a bale per acre. Receiving higher wages and larger crop shares, some Black workers in the Southern bottomlands managed to save enough money to buy some land. The dream of land ownership impelled "quite a number" of Black families to buy tracts of unimproved woodlands from railroad companies and White planters. Purchasing uncleared land on credit, they lived and labored on nearby plantations, clearing their own land when time permitted. If they were lucky, families improved their land, eventually becoming independent farmers.[54]

One Mississippi planter, who was writing around the turn of the twentieth century, claimed that it was impossible to travel across the Mississippi Delta "without observing Negro [*sic*] landowners everywhere." Black farmers were "scattered over its entire area, holding tracts varying in size from a town lot to more than a thousand acres."[55] There was much truth to this observation. By 1900, two thirds of the farm owners in the Mississippi Delta were Black.[56]

Although numbers of Black settlers managed to climb the agricultural ladder to land ownership, most remained landless laborers. Each fall, labor agents recruited thousands of Southern Blacks to pick cotton in the bottomlands. Viewing the rich soils and abundant crops, many decided to stay, becoming laborers, croppers, and renters on plantations.[57] Most of these Black newcomers remained tenant farmers, who drifted from plantation to plantation. As an example, a Mississippi planter, who began operating Dunleith plantation in 1899, recruited fifty-eight Black tenant families to work 1,064 acres of cotton land. By 1904, only eight of the original tenant families were still left on Dunleith, and none of the families who left the plantation ever acquired land.[58] Trapped in a cycle of poverty, many Black newcomers discovered the harsh reality of life in the Southern bottomlands. Despite its inviting fields, the Mississippi Valley was "not a democratic frontier eager for settlers" but was "a plantation country hungry for labor."[59]

Although Mississippi, Louisiana, Arkansas, and Missouri welcomed the influx of Black workers during the late nineteenth century, these states also hoped to attract Northern settlers and foreign laborers. These states published immigrants' handbooks and farmers' guides for prospective settlers, and these publications glowingly described the fertile soils, healthy climates, and friendly inhabitants of the Southern bottomlands.[60] In addition to these laudatory state publications, railroad companies also touted the fertile Mississippi Valley, hoping to sell acreage to Northern land seekers. As an example, the Illinois Central corporation, which controlled the Y&MV railroad, was particularly anxious to attract Northern settlers to the bottomlands. Holding more than five hundred thousand acres of Y&MV railroad land in the Mississippi Delta, the Illinois Central planned to sell tracts on time to Northern farmers, who would not only buy railroad lands but who would also generate business by shipping produce to market. But despite corporate and state efforts, there was no rush of Northern land seekers to the Southern bottomlands. Fearing fevers and floods, most Northern land seekers sought farmsteads in the Western prairielands and not in the Southern swamplands.[61] Yet, the flood of publicity lured a few Northerners and foreigners to the Southern bottomlands. Some midwesterners settled along the railroads in the Mississippi Delta, where they gradually adopted Southern-style farming. In addition, some German-American farmers from the Midwestern states located on the Grand Prairie of eastern Arkansas. Finding a prairie environment with little malaria, these Teutonic midwesterners pursued livestock farming on grasslands that had been overlooked by Southern farmers.[62] Unlike the midwesterners who settled on the Grand Prairie, the Italian laborers, who worked on Sunnyside plantation in Chicot County, Arkansas, found nothing but malaria, poverty, and indebtedness. Many of these workers abandoned Sunnyside, and they moved with Father Pietro Bandini to establish the Tontitown colony in northwestern Arkansas.[63] Attempts to entice Chinese laborers to the bottomlands also came to naught. Finding disease and indebtedness on bottomland plantations, these Asian workers drifted off to more favorable areas.[64] Thus, despite corporate and state efforts to portray the Southern bottomlands as a healthful and fruitful place, few foreign laborers or Northern farmers settled in the forbidding swamplands.

Northern timbermen, however, found the Mississippi Valley a profitable field for investment. During the late nineteenth century, timber companies invaded the Southern bottomlands, purchasing swampland forests that contained marketable hardwood trees. Railroad corporations, state governments, levee districts, and land speculators gladly sold extensive tracts to timber companies for a few dollars an acre. As an example, the George T. Houston and Company of Chicago, Illinois, purchased 150,000 acres of Delta timberland from the Y&MV railroad for $1 million. Cutting the marketable cypress, gum, oak and cottonwood trees from bottomland tracts, timber companies processed the timber in sawmills, stave shops, and box factories, which sprang up along the railroads.[65]

These timber factories and sawmills, which offered good wages for the time, attracted poor White job seekers from the Southern hillcountry. In addition to the poor Whites who sought work in the timber industry, the bottomlands also lured Southern White land seekers, who planned to carve farms from cut-over timberlands. Laboring on lonely farmsteads and working in sawmill towns, these Southern White newcomers often improved their economic status, but they also battled diseases and floods.[66]

Even acclimated White Southerners had difficulty in adapting to a "swamp country." Residing on their 140-acre farm in Desha County, Arkansas, Nannie and William Jackson endured a life of hard work and frequent illness. Keeping a diary during the years between 1890 and 1891, Nannie noted that on July 25, 1890, William bought "a bottle of quinine" to ease her malarial chills. On March 31, 1891, Nannie's daughter lay in bed with a high fever, and Nannie dosed her with niter and calomel. A few days later, a spring flood inundated the Jackson's farm, nearly entering their house.[67] Given the threat of floods and fevers, even White Southerners looked askance at the fertile but forbidding bottomlands.

Despite a well-deserved reputation as a swampy and sickly place, the Southern bottomlands managed to attract thousands of settlers between 1880 and 1900. The building of levees and railroads permitted the settlement of flooded and isolated swamplands, which had been avoided by settlers before 1880. Writing in 1893, William Stovall of Coahoma County, Mississippi, marveled at the changes wrought by improved flood protection and transportation:

> Building the levees along the Mississippi River and protecting the country from overflow has made it possible to put in cultivation large areas that could not be utilized without such protection. It has brought about a large increase in our population, mostly laborers, and a vast amount of "new ground" has been cleared up and brought into use, and thus has been added to the . . . large quantity of virgin soil of the finest quality. The other factor is the building of the Louisville, New Orleans and Texas Railroad, which passes through the delta [*sic*]. All along its line, plantations have been opened up, towns built, and in this way there has been great development.[68]

Thanks to this "great development" during the late nineteenth century, the population of the bottomland counties and parishes soared from 390,014 (see Table 1) to 648,322 between 1880 and 1900. Despite this impressive population growth, areas with better flood protection and more reliable transportation attracted more settlers than other areas. As an example, the Mississippi Delta, which possessed stronger levees and better rail service, added more residents than the Louisiana Delta, which had weaker levees and sketchier rail service. Between 1880 and 1900, the ten counties that lay within the Mississippi Delta added 105,348 Black and White residents, but the eight parishes that lay within the Louisiana Delta gained only 5,963 Black and White inhabitants. By 1900, the Louisiana Delta had a population density of only twenty people per square mile, whereas the Mississippi Delta had a density of thirty-six persons per square mile. Throughout the Southern bottomlands, the population density averaged twenty-seven people per square mile by 1900.[69] Given the increasing population density, the Mississippi Valley bottomlands were becoming "intensive agricultural frontiers" by the turn of the twentieth century. Population densities approaching forty-five people per square mile indicated "a highly successful agriculture," based on more intensive agricultural practices.[70]

The increasing population in the Southern bottomlands prompted agriculturalists to intensify their livestock-raising practices. When the bottomlands were extensive agricultural frontiers with low population densities, agriculturalists fenced in their fields, and they allowed their livestock to wander on the unfenced lands, or "open range." But as the Southern bottomlands evolved into intensive agricultural frontiers with higher population densities, agriculturalists began to fence in their livestock, and they left their fields unfenced.

Before the Civil War, livestock raising was so important that Southern state legislatures required farmers to enclose fields with fences so livestock were free to range on the unfenced lands. Agriculturalists typically enclosed their fields with "worm fences," which were zig-zag stacks of split rails. Although well-built rail fences kept out cattle and hogs, these wooden enclosures consumed much labor. Yet, before the war, labor for fence building was available during the winter months between the fall harvest and the spring planting. Farmers put their slaves to work building rail fences during the slack winter months, when time was available to fell trees, split rails, and construct enclosures. But after the Civil War, it proved difficult to hire free workers to split rails and make fences. The cost of making rail fences with free labor became an expensive burden for crop growers. Seeking to escape this burden, crop growers petitioned state legislators to change the fence laws. Petitioners wanted legislators to pass "stock laws" that required stock owners to enclose cattle and hogs on their own property. By requiring stock owners to enclose their animals, these laws allowed crop growers to leave their fields unfenced, thus shifting the cost of fencing from crop growers to stock raisers.[71]

The effort to pass such stock laws began as early as 1866 in Mississippi. In that year, the Mississippi state legislature allowed voters to petition county courts in order to enclose livestock, so that planters could cultivate their crops without fences. By 1880, more than thirty densely settled, crop-growing counties in Mississippi required stock owners to enclose their animals, so crop growers did not have to fence their fields. But within the sparsely settled Mississippi Delta, only Tunica County had adopted a stock law by 1880. In the remaining Delta counties, the open range prevailed, and scrub livestock still wandered on the unfenced lands.[72] The campaign to enclose livestock, however, accelerated after the introduction of wire fencing. In 1879, barbed-wire fencing, which had been developed in the Western prairie states, became available in the Southern states. Barbed-wire strands, which were nailed to fence posts, proved sharp enough to keep in the most determined cattle. And in 1883, the introduction of woven-wire fences offered a means of enclosing hogs. Fashioned from intersecting wires, woven-wire fencing, which was nailed to wooden posts, kept in the most restless hogs. Woven-wire and barbed-wire fences not only effectively enclosed livestock, but also saved money and labor. It was cheaper and easier to put up wire fences than to split logs, stack rails, and build worm fences. Wire fences even lasted twice as long as split-rail fences. As wire fencing supplanted split-rail fences, the resistance to stock laws began to crumble.[73] To ease the adoption of stock laws, the state of Mississippi passed a local option law in 1892, allowing voters within precincts to petition county courts for elections in order to enclose livestock. If a majority of voters agreed, precincts enclosed livestock, so crop growers were free to leave their fields unfenced. After the passage of this law, county precincts across the Mississippi Delta began closing the open range.[74]

After closing the range, it was possible to improve livestock herds by practicing selective breeding. Fenced enclosures allowed stock owners to breed purebred bulls and boars with scrub cows and sows. Breeding scrub stock with purebred animals yielded mixed-breed offspring that possessed the hardiness of their scrub mothers and the meatiness of their purebred sires. Yet, whenever stock owners introduced purebred sires to mate with scrub dams, these purebred animals soon succumbed to "tick fever" or "hog cholera." Although scrub animals possessed a degree of immunity to these endemic stock diseases, purebred animals proved vulnerable to these maladies. Thus, scrub stock could not be improved until tick fever and hog cholera were eliminated so purebred stock could be safely introduced.[75] These stock diseases, however, would not be conquered until the early twentieth century (see chapter 3), and livestock improvement in the Southern bottomlands would have to await the advent of a new century.

3

The Southern Bottomlands During the Early Twentieth Century (1900–1920)

> For sixty years, these lands have been partially protected from overflow, but until within the past twenty years, the levees were crude, indeed they were little more than artificial mud banks; but today, through government appropriations and government supervision, assisted by the State Levee Boards, a system of levees has been established which protects the Delta lands except during years of great floods, and the soil is so very fertile that even when there comes a year of overflow, as soon as the water recedes the farmer puts in his crop with good prospects.[1]

Before 1900, settlement of the Southern bottomlands proceeded in a limited fashion. Settlers located on frontlands and on ridgelands, clearing the timber from new fields. To protect their frontland fields from flooding, they built levee embankments, which held back floodwaters in most years. Hugging the frontlands and ridgelands, bottomland residents left thousands of acres of rich backlands untouched because these swampy areas required costly drainage. But after 1900, settlement proceeded in a more systematic fashion. Strong levee embankments protected the bottoms from all but the worst floods. In turn, drainage projects permitted the reclamation of swampy backlands. Finally, the cutting and clearing of swampland forests permitted the cultivation of these drained backlands. Thanks to levee building, land draining, and land clearing, the agricultural potential of the Southern bottomlands was finally realized in the early twentieth century.

Flood protection had made great strides since the Civil War. By 1897, state levee districts lined the lower Mississippi Valley (see chapter 2). Following the great 1897 flood, these districts rebuilt the levee line. But in 1903, floodwaters overlapped the rebuilt levees. Although the 1903 flood was less destructive than the 1897 deluge, the levee districts raised the tops of the embankments above the 1903 flood levels. Between 1903 and 1912, no strong floods tested the levee line, and confidence in the levee system soared during these years.[2]

Given the improved flood protection, the Southern bottomlands attracted the attention of land speculators, timber entrepreneurs, and prospective planters. These investors were eager to purchase unimproved swamplands that were valued at less than ten dollars an acre. Yet, this unimproved swampland had hidden costs. If timber companies purchased swampland forests and removed the marketable trees, they were left with unproductive cut-over lands, which became a tax burden on the companies. Hoping to escape burdensome land taxes, timber companies were willing to sell cut-over lands to agriculturalists for a few dollars an acre. Farmers, however, not only had to finish clearing these cut-over lands, but they also had to ditch and drain these lands before cultivation. Although land could be cleared for as little as ten dollars an acre, the cost of digging ditches to drain land was more than fifteen dollars an acre. Given these hidden drainage costs, the price of draining swamplands was often prohibitive for private landowners. In order to drain the cut-over and timbered lands of the Southern bottomlands, it was necessary for landowners to pool their resources by creating "drainage districts," which would dig the expensive ditches needed to drain swamplands.[3]

Drawing on earlier Midwestern drainage laws, Southern state legislatures passed laws that permitted the creation of drainage districts in the lower Mississippi Valley. Drainage laws allowed landowners in an area to petition for creation of a drainage district. After engineers examined the lands to determine if drainage was feasible, the petition was approved, surveys were completed, and district commissioners were appointed to oversee the drainage work.[4] Because drainage districts usually enclosed tracts of unimproved swamps, the draining of timbered swampland was a formidable task. It was virtually impossible for shovel-wielding men or mule-drawn scrapers to excavate the massive ditches needed to drain swamplands. In order to excavate drainage ditches, district commissioners contracted with engineering firms that operated steam-powered floating dredges. Floating dredges, which were equipped with dipper buckets, made it possible to excavate ditches through the timbered swamplands. After engineers surveyed the right-of-way, workers cleared the way by felling trees, removing logs, and shattering stumps with dynamite. Following the removal of the timber cover, dredges excavated the debris and muck. Using yard-wide dipper buckets, floating dredges excavated up to one-thousand cubic-yards per day. Although it cost as much as three thousand dollars per mile to excavate ditches with dredges, it was the only feasible mode of draining swamplands. To pay for costly drainage work, district commissioners sold bonds to investors and imposed drainage taxes on local landowners[5] Because these drainage taxes added to the cost of operating plantations and harvesting timber, this tax burden often chilled the enthusiasm of planters and timbermen.

The fear of drainage taxes often stirred local opposition to drainage projects. As an example, the planters who lived on the well-drained frontlands along the Mississippi River, resented the high taxes needed to fund the drainage of the swampy backlands in the huge Tallahatchie Drainage District, which covered the northern third of the Mississippi Delta. Leading the opposition to the district, these parochial riverside planters forced the Mississippi state legislature to repeal the district in 1912—only four years after its creation.[6]

There was also parochial opposition to drainage projects in eastern Arkansas. When pro-drainage forces petitioned to create Drainage District No. 9 in Mississippi County, Arkansas, in 1909, the local planters, who feared high property taxes, and the lumber companies, which hoped to avoid drainage taxes, opposed the petition. Despite this opposition, drainage advocates eventually prevailed. The drainage district was approved, and drainage commissioners were chosen. Selling drainage bonds to Northern investors, the district commissioners contracted with engineering companies to complete the work. Using twenty dredges, contractors dug more than three-hundred miles of drainage ditches. When work began in 1911, Drainage District No. 9 enclosed 193,000 acres of land, but only 52,000 acres were improved farmland. By 1926, the district contained 131,000 acres of improved farmland.[7]

Drainage proponents also prevailed in southeastern Missouri, where they created one of the largest drainage districts in the United States. Begun in 1907, the Little River Drainage District encompassed 488,000 acres in seven counties. When drainage work began, there were only 22,000 improved acres within the Little River district. By 1926, the district contained 265,000 improved acres of farmland.[8]

Despite the failure of the huge Tallahatchie Drainage District, drainage advocates in the Mississippi Delta created dozens of smaller drainage districts. Typical of these smaller Delta districts was the Belzoni drainage project, which lay between the Sunflower and Yazoo rivers. Created in 1911, the Belzoni district enclosed 91,000 acres—the bulk of which was unimproved swampland. Because the natural streams that drained the Belzoni swamps were jammed with fallen trees, logs and brush, this debris prevented the drainage of the swampland, which was submerged by floodwaters for most of the year. Not only was the Belzoni swampland subject to local flooding, but floodwaters from the Mississippi River backed up the Yazoo River and inundated the Belzoni swamps. In order to drain the flooded Belzoni swamps, engineers planned to clear the natural drainage channels, excavate drainage ditches, and build levees along the west banks of the Yazoo River to keep out floodwaters. When work began on the Belzoni project, there were only 20,000 improved acres within the district. By 1926, the Belzoni district contained 44,000 improved acres of farmland.[9]

Because drainage districts often succeeded in reclaiming thousands of acres of fertile swamplands, the drainage movement spurred the settlement of the Southern bottomlands during the early twentieth century. Drainage projects often worked wonders in Southern bottomland counties. This was certainly the case in Poinsett County, Arkansas, during the early twentieth century. Although Poinsett included prairies and ridgelands, the eastern part of the county lay in the swampy St. Francis River Valley. Hoping to reclaim this swampland, the business community in Marked Tree, a town on the St. Francis River, petitioned for the creation of drainage districts. As drainage work progressed, lumber companies removed the marketable timber from the drained swamplands, and businessmen bought the cut-over lands. Because cotton prices soared after 1900, land buyers planned to transform the cut-over lands into cotton plantations. To accomplish this transformation, businessmen-planters recruited White and Black tenant farmers in order to clear and plant the land. Although White tenants suffered from malaria, and Black tenants suffered from racial violence, these pioneering farmers carved cotton and corn fields from the drained swamplands.[10]

A Northern traveler, who was passing through eastern Arkansas in the early 1900s, witnessed the transformation of forests into fields. Stopping at a sawmill village along the railway, the visitor saw several small farms, which had been hewn from cut-over lands. Walking through the watery woods, he came on "a big cornfield" that was filled with "withered last year's stalks." A farmer, who was clearing off the old stalks, told the visitor that he did "not own the land." Instead, he paid "one-third of the corn" and "one-fourth of the cotton" that he raised as rent to the landowner. Leaving the tenant farmer, the traveler followed a logging railroad, which used mule-drawn cars to haul logs to the sawmill. Entering a timber camp, he found a cluster of frame houses with canvas roofs. These temporary shelters, which were easily moved every few months, housed the wood cutters and their families. Cutting out the marketable trees, timbermen pushed deeper into the swampland forests, leaving cut-over lands strewn "with a chaotic mass of brush and limbs," in their wake.[11]

Because timber companies owned much of the land within drainage districts, these corporations gladly sold cut-over lands in order to avoid drainage and property taxes. Consequently, timber companies established land agencies to market cut-over swamplands. Advertising widely in the Midwestern states, agencies offered to sell tracts on generous terms to farmers and speculators. Thanks to aggressive advertising, timber companies managed to sell "a large amount of land" to "northern [*sic*] purchasers, many of whom settled upon their new property and made good farms." These Midwestern settlers were particularly numerous in northeastern Arkansas and southeastern Missouri, where they raised grain as well as cotton. Despite this influx of midwesterners, timber companies sold more cut-over land to "farmers and townspeople living in and near the [drainage] districts" in Missouri, Arkansas, and Mississippi. Investing in cut-over lands, Southern speculators and agriculturalists planned to develop plantations and arms in the cut-over swamplands. These local land buyers were soon joined by a stream of White and Black Southerners from the uplands of Missouri, Arkansas, Kentucky, Tennessee, and Mississippi. Leaving their worn-out hill farms, these settlers hoped to till the "drained alluvial soils" of the reclaimed bottomlands. Although most of these hill-country newcomers became tenant farmers on bottomland plantations, some later became landowners.[12] Selling lands to Southern and Northern settlers and speculators, timber companies disposed of much of their cut-over swampland within drainage districts.

Despite numerous success stories, drainage districts were costly undertakings. Because drainage work was so expensive, districts issued interest-bearing bonds to fund the work, and they assessed drainage taxes on acreage in order to pay the interest. In the case of Drainage District No. 9 in Mississippi County, Arkansas, it cost $2,990,000, or about $15.50 per acre, to reclaim the district's swamplands. The drainage taxes on these reclaimed swamplands were a modest $1.22 an acre each year, but there were also additional local and state taxes of $2.28 per acre per annum. Totaling $3.50 an acre each year, this tax burden could be borne by agriculturalists, who were raising crops on this acreage, but it could not be borne by timber companies, which were holding cut-over lands, or by land speculators, who were hoarding unsold acreage. Because corporations and speculators often balked at paying drainage taxes, many districts struggled to pay the interest on drainage bonds. By the 1920s, there were strident calls for federal aid to save indebted drainage districts from default.[13]

Although many drainage districts plunged deeply into debt, and although some districts failed to live up to overblown expectations, the drainage movement reclaimed much of the alluvial swampland in southeastern Missouri, eastern Arkansas, and northwestern Mississippi.[14] Swampland drainage, however, made far less progress in northeastern Louisiana. In the state of Louisiana, drainage work focused on the alluvial marshlands of southern Louisiana. If drained, these treeless marshlands promised to yield bountiful crops of sugar cane and rice. Given the attractiveness of the Louisiana marshlands, drainage advocates overlooked the timbered swamplands of the Ouachita–Tensas basin in northeastern Louisiana.[15] In 1910, when A. E. Morgan and Associates surveyed the Fifth Louisiana Levee District, which lay between the Tensas and Mississippi rivers in northeastern Louisiana, they found that drainage work had barely begun. Planters and farmers cultivated the higher ground along the rivers, but the interior swamplands were simply too wet to cultivate. The Fifth Levee District contained 1,263,000 acres of uncultivated swampland but only 299,000 acres of improved farmland. In Concordia Parish, which was bounded by the Mississippi, Red, Black, and Tensas rivers, the only cultivated land was on the riverside frontlands, and the rest of the parish was swampland "waste," which "overflowed during high water" to "depths varying from a few feet to 20 feet or more." To reclaim the swamplands of the Fifth Levee District, Morgan and associates recommended clearing the natural channels to foster drainage, leveeing the streams to exclude floodwaters, and excavating huge ditches to drain swampland. The estimated cost of these drainage works was nearly eleven million dollars.[16] Given the huge cost of swampland reclamation in northeastern Louisiana, local residents launched only small-scale drainage projects in the years before 1930. Drainage districts failed to reclaim the vast swamplands of northeastern Louisiana, and parishes such as Concordia actually lost residents during the early twentieth century.[17]

Although the drainage movement stalled in northeastern Louisiana, drainage districts opened thousands of acres to agricultural settlement in Mississippi, Arkansas, and Missouri. As arable lands were reclaimed from swamps, settlers in the bottomlands faced the tedious task of clearing cut-over land and "raw swampland." In the case of cut-over acreage, "the land had to be cleared of small trees, brush, and such odd logs and tree tops as the lumber crews had left, before a crop could be planted." Typically, this was done by slashing the undergrowth, piling the brush, and burning the piles. After firing the slash, farmers planted corn and cotton, "among the stumps which would be mostly gone after about five years of burning, rotting and plowing." Because it was possible to plow around stumps, there was no need to waste labor by excavating these obstructions.[18] Despite the inconvenience caused by stumps, it was easier to clear cut-over land than to remove the timber cover from raw swamplands. Yet, if agriculturalists were clearing uncut timberland, it was possible to sell the marketable timber from that acreage. If sawmills were convenient, settlers cut and sold the best timber, leaving the rest of the trees to be "girdled" and "deadened." Cultivating around the dead trees, farmers gradually removed these obstructions from their fields.[19] By deadening trees, slashing brush and burning debris, settlers slowly claimed fields from raw and cut-over swamplands.

Pioneer farmers in the Southern bottomlands frequently cleared their own land, but planters and speculators employed others to clear swampland acreage.[20] In the Mississippi Delta, a cheap method of bringing "new land" into cultivation was to offer wooded tracts "rent free" to Black tenants "for three years." The tenants kept all the cotton and corn that they raised, while they cleared off the timber and brought "the soil under cultivation."[21] But in the Missouri Delta, landowners hired White families from the upland South to clear land by the acre. Arriving in southeastern Missouri with their "axes and crosscut saws," families from the hills of Missouri, Arkansas, Tennessee, and Kentucky agreed to clear land for eight to twelve dollars an acre, depending on the density of the forest cover. Landowners assigned ten-acre strips of woodland to each family and provided them with a "rough shack" for a residence. Clearing timber from the swampland, many of these "valiant axemen and axewomen" later became tenants on the acreage that they cleared.[22]

Clearing swamplands in the lower Mississippi Valley, workers, tenants, and farmers were "dead sure to be smitten with malaria, ague, chills and fever, or whatever you choose to call this wretched mosquito-borne malady."[23] The swamps of the Southern bottomlands harbored anopheline mosquitoes, which were vectors for the malaria plasmodia. Although scientists discovered the relation between mosquitoes and malaria by turn of the twentieth century, this knowledge diffused slowly to the lay public.[24] During the early twentieth century, many rural Southerners still believed that malaria was "caused by an imaginary and indefinable miasma" in the swamps, and they regarded mosquitoes as only troublesome pests.[25] Despite misconceptions about the origins of malaria, wealthier Southerners had long taken quinine to ward off malarial chills. This efficacious drug, however, was too expensive for the poorer people, who relied on home remedies to ease the ague (see chapters 1 and 2). Although the cost of quinine dropped during the nineteenth century, it retailed for nearly two dollars an ounce during the early twentieth century.[26] Unable to afford quinine when they moved to the bottomlands, many poorer people turned to cheap patent medicines, which promised to cure malarial chills. Local merchants sold "millions of bottles" of "chill tonics" to unsuspecting newcomers. Dosing themselves with patent nostrums, these "ague-shaken swampers" continued to suffer from malaria.[27] Fortunately, drainage work and land clearing eased the threat of malaria by removing sluggish swampwaters, where anopheline mosquitoes bred.[28] In addition, better-educated swampland residents, who learned of the relation between mosquitoes and malaria, screened their houses, protected beds with mosquito nets, and lit smudge fires to ward off mosquitoes.[29] Bottomland residents who took these precautions contracted malaria less frequently than poorly educated workers and tenants, who lived in rickety houses that lacked screens and nets. Enduring the bites of voracious mosquitoes, bottomland workers and tenants sickened and sometimes died from malaria.[30]

Whether swampland residents recognized mosquitoes as malarial vectors or as annoying vermin, they all realized the threat that insect pests posed to bottomland crops. The wet conditions in the bottomlands allowed insects to proliferate, and monocropping abetted the proliferation of pests. In the bottomlands, two thirds of the cultivated land was planted in cotton, and most of the remaining acreage was cropped in corn. A common crop rotation was to plant cotton in a field, follow this with a crop of corn (mixed with cowpeas), and then plant cotton again. Cotton, of course, was the money crop, corn fed the work stock, and the leguminous cowpeas added nitrogen to the soil.[31] This simple rotation provided cash, fodder and fertilizer, but the extensive fields of cotton and corn invited insects to feast on these crops. Several species of moths laid their eggs on the growing plants, allowing their larvae ("cotton worms," "boll worms," "corn borers" and "cut worms") to feed on leaves, fruits, and stalks. The abundance of these insect pests, nevertheless, varied yearly. In those years, when rains saturated the fields or when floodwaters inundated the lands, plagues of these pests devastated crops, forcing agriculturalists to replant their fields. In some cases, it was necessary to replant two times before sufficient seedlings survived the insect onslaught. Although moth larvae were a periodic pestilence in the bottomlands, these pests proved far less destructive than a tiny weevil that attacked cotton flowers, squares and bolls.[32]

Dubbed the "cotton boll weevil," this pest crossed the Rio Grande in 1892 after infesting Mexico's cotton crops. Invading southern Texas, weevils found extensive cotton fields to infest. Female weevils laid their eggs in cotton flowers, the growing grubs consumed the squares, and adult weevils fed on young cotton bolls. Because it took only two to three weeks for an egg to develop into an adult, generations of weevils quickly overran cotton fields. Flying with the winds, weevils moved eastward at a pace of more than sixty miles a year, and they overran much of Texas by the turn of the twentieth century. Writing in 1901, W. D. Hunter, an entomologist with the U.S. Department of Agriculture, predicted that the boll weevil would be "a permanent factor in cotton culture" and that weevils would soon "spread outside Texas."[33] He was correct on both counts. Leaving Texas after 1900, weevils threatened the cotton fields of the Mississippi Valley. Before 1910, weevils invaded northeastern Louisiana, southeastern Arkansas, and northwestern Mississippi. After 1920, weevils even reached northeastern Arkansas and southeastern Missouri. As weevils invaded the Mississippi Valley, cotton yields began to plummet. In newly infested areas, yield losses of 50 percent or more were not uncommon.[34]

Although weevils seemed indestructible to cotton growers, who tried to poison them with everything from kerosene to "Paris green" (acetate and arsenate of copper), entomologists discovered some chinks in the weevils' armor. When fall frosts killed the cotton plants, the adult weevils hibernated in the dead stalks, leaf litter, and other trash in the fields. With the return of warm weather, the surviving weevils were poised to attack the next cotton crop. Because weevils were vulnerable during the winter months, scientists advised farmers to pull up dead stalks and burn the trash after they finished picking cotton. Some weevils survived this cleanup, but it bought crucial time for cotton growers who planted, cultivated, and picked the cotton plants before weevil populations proliferated. To further reduce weevil damage, they urged farmers to raise fast-maturing cotton varieties that fruited before weevils became too numerous. They also recommended the use of commercial fertilizers to hasten cotton maturation. By following these recommendations, cotton farmers reduced weevil damage to more tolerable levels, and they managed to successfully raise cotton under "boll-weevil conditions."[35]

Entomologists not only developed techniques for reducing weevil damage, but they also discovered that the severity of weevil infestation depended on local conditions. They found that weevils thrived in areas with mild winters, wet soils, and timber cover. Weevils were far less numerous in regions with cold winters, dry soils, and open country.[36]

Because northeastern Louisiana had mild winters, wet soils and timber cover, this area proved an ideal habitat for the boll weevils that ravaged the cotton plants in riverside fields. Cotton production in the northeastern parishes plunged to postbellum lows. In 1900, Concordia Parish harvested 27,799 cotton bales. Ten years later, the parish gathered only 4,929 bales. And by 1920, the parish cropped only 3,591 bales. The two cottonseed mills in Vidalia, which had crushed local seed to produce oil, shut their doors, due to the lack of cottonseed. Some desperate Louisiana planters abandoned cotton, and they embraced rice—a crop that had been introduced to the Louisiana Delta by the Naylor family. Learning to raise rice in southwestern Louisiana, the Naylors moved to Madison Parish in 1909, introducing this crop to northeastern Louisiana. Discovering that rice grew in wet fields along swamplands, some planters successfully raised this grain for market. As early as 1910, a rice mill was opened in Vidalia to process the local crop. Many Concordia planters, however, refused to give up cotton, and they continued to raise this crop in traditional fashion. As late as 1920, some still planted slow-maturing cottons, they used no fertilizers to hasten growth, and they took few measures to control the weevils that infested riverside fields.[37]

Cotton planters in the Mississippi Delta, however, proved more adaptable to the weevil menace. Even before boll weevils reached the Delta, planters subscribed money in 1904 to buy land for a state agricultural experiment station at Stoneville. Because Delta planters raised slow-maturing fine-staple cottons, these late-fruiting plants were highly vulnerable to weevils. Consequently, agronomists at Stoneville began breeding quick-maturing cottons that fruited before weevils overran fields. These experiments proved successful, resulting in such fast-fruiting types as "Stoneville" and "Delfos." In addition to developing fast-maturing cottons, Stoneville agronomists urged Deltans to diversify their crops in order to reduce weevil damage.[38] They recommended that planters rotate cotton crops with legumes (cowpeas, soybeans, alfalfa and vetch), since the "legumes cultivated in the South" were "not greatly subject to insect attack."[39] Bracing for the weevil invasion, which arrived in 1908, Delta planters were relieved to find that weevils proved less troublesome in northwestern Mississippi than in northeastern Louisiana. In the Mississippi Delta, colder winters held the weevils in check. Despite the arrival of weevils, cotton production continued to rise.[40] But in years with mild winters, weevils proliferated in the Mississippi Delta. In those years, Deltans prayed for an effective poison to combat weevils. Because weevil grubs were sheltered by cotton flowers and squares, they proved impervious to conventional poisons. Adult weevils, however, proved more vulnerable to pesticides. In 1918, B. R. Coad, an entomologist with the U.S. Department of Agriculture, found that calcium arsenate proved effective against adult weevils. When dusted on cotton plants at night, calcium arsenate poisoned the dew that adult weevils sipped from leaves. In order to dust cotton plants with calcium arsenate, scientists devised hand-turned blowers as well as mule-drawn dusters. They even experimented with airplanes to dust cotton fields. Because calcium arsenate proved effective, Delta planters purchased tons of this weevilcide.[41] By using effective poisons and by raising fast-maturing cottons, Delta cotton growers survived the weevil onslaught.

Boll weevils also invaded eastern Arkansas after the turn of the century, disrupting the state's cotton industry. Weevils overran southeastern Arkansas by 1910, but northeastern Arkansas escaped the invasion until 1920. As weevils entered Arkansas, agronomists urged cotton growers to diversify their crops. They condemned the "common system" of planting two thirds of the tilled land in cotton—a practice that invited weevil infestation.[42] Agricultural reformers called on farmers to plant more weevil-resistant crops such as alfalfa, cowpeas, and soybeans. These legumes not only enriched soils with nitrogen, but they also furnished fodder for livestock. Although legumes held great promise as field crops, rice proved to be "the most profitable" weevil-resistant crop in eastern Arkansas.[43]

Arkansans raised a little swampland rice during the late nineteenth century, but transplanted Midwesterners introduced prairie rice cultivation to the Grand Prairie of eastern Arkansas during the early twentieth century. The Grand Prairie, an extensive grassland, lay between the White and Arkansas river valleys (see chapters 1 and 2). Because this huge prairie was underlain by impervious clay subsoils which retained water, it was difficult to grow cotton and corn on the water-logged soils. The prairie, however, yielded an abundance of native grasses, which fed cattle for ten months a year. Noting the abundance of grass, Midwestern cattle farmers settled on the Grand Prairie after the Civil War. Attracting British- and German-Americans from the Midwestern states, the Grand Prairie became an island of Midwesterners living amidst a sea of Southerners.[44] Hearing of the Grand Prairie, the John Morris family of Nebraska moved to eastern Arkansas in 1894. Grazing cattle on the Grand Prairie, John Morris encouraged his brother-in-law, William Fuller, to leave Ohio and join him in Arkansas. After moving to Arkansas, Fuller went on a hunting trip to the Gulf of Mexico in 1896, where he saw the rice fields of the southwestern Louisiana prairies. Returning to Arkansas with a sack of rice, he tried to grow the grain on the Grand Prairie. Because his experiment failed, Fuller returned to Louisiana, where he hoped to learn the secrets of rice growing. Renting a farm near Jennings, Louisiana, in 1898, Fuller learned how to irrigate rice fields, using water from deep wells to flood crops during the growing season. Returning to Arkansas in 1904, Fuller was determined to raise rice in his home state. His neighbors offered him a one-thousand dollar bounty if he could raise rice on the Prairie, so Fuller drilled a deep well, installed a pump, and irrigated seventy acres of prairie rice. Harvesting more than five thousand bushels from his rice field, Fuller earned the bounty and even turned a profit on his crop. After Fuller's experiment succeeded, the Arkansas agricultural experiment station at Lonoke began sowing prairie rice, thus proving the commercial value of this crop. By 1906, Arkansans sowed five thousand acres of prairie rice, and rice fields dotted the Grand Prairie.[45]

The Grand Prairie proved an ideal environment for rice growing because the impervious clay subsoils held irrigation water during the growing season. After plowing the prairie topsoil, rice growers seeded the land with animal-drawn drills. Using plows, farmers then created levees to hold irrigation water, and they employed steam-or gasoline-powered pumps to flood rice fields with water from deep wells. When rice plants matured, farmers opened the levees, drew off the water, and used animal-drawn machines to harvest the rice. Some rice farmers even employed tractors to plow land, drill seed, and harvest crops.[46] After harvesting, growers sent their grain to mills for processing. During the first years, they shipped rice by rail to distant mills in New Orleans, Louisiana. But in 1906, local entrepreneurs established a rice mill in the town of Stuttgart on the Grand Prairie. By 1910, four more mills were processing prairie rice from local fields. Farmers hauled sacks of rice to mills in animal-drawn wagons or in new motor trucks. Using trucks, tractors, and machines, Arkansas rice growers were among the most technologically sophisticated farmers in the American South.[47]

Although prairie rice required a heavy investment in technology, the crop saved labor. It required less than five man days of labor to raise, harvest, and market the rice from one acre of land in eastern Arkansas. In contrast, it took twelve man days to raise, pick, and market the cotton from one acre in eastern Arkansas.[48] Rice not only saved labor, but it was also a familiar crop to mechanized grain farmers from the Midwestern states. Midwestern grain farmers had long used machines to seed and harvest wheat and other small grains. Not surprisingly, the "most successful rice farmers in Arkansas" were "former wheat growers from the northern [*sic*] and middle states."[49] Discouraged by low wheat prices, many Midwestern grain farmers moved to eastern Arkansas, where they planned to raise rice on the prairie land. Hoping to attract Midwesterners, railroad companies and land speculators organized railway excursions to eastern Arkansas. Showing off the prosperous rice farms, corporations and developers sold prairie lands on terms to prospective rice farmers.[50] By 1910, unimproved prairie land was selling for twelve to thirty-five dollars an acre, and improved rice lands were fetching forty to one-hundred dollars per acre. Given the rising land values, many Midwestern newcomers found it wiser to rent rather than to buy acreage. Renting allowed farmers "with small capital" an opportunity to enter the rice-farming business and then earn money for machinery and land. Under the usual rental arrangement in Arkansas, landlords furnished the land, seed, water, stock, and machinery, while the renters sowed, tended, and harvested the rice crop. Paying half of the crop as rent, tenants kept the rest of the rice. After renting lands for several years, the newcomers often saved enough money to become landowning rice farmers. Although it required more than five-thousand dollars to obtain the land and equipment for a rice farm in eastern Arkansas, rice farming was quite profitable. Rice yielded fifty to one hundred bushels per acre, the cost of raising the grain ranged from sixteen to thirty-two dollars an acre, and rice sold for one dollar a bushel in 1915. Encouraged by sizable profits, Arkansans were raising more than one hundred thousand acres of rice by that year.[51]

The rice boom touched off a land rush in east-central Arkansas. Rice farmers occupied the Grand Prairie and claimed the smaller prairies in the White River Valley. Noting that the timbered bottomlands of the White River Valley possessed the same clay subsoils as the prairies, farmers began tentatively clearing these clay-bottomed swamplands. Yet, clearing timberland to create rice fields proved an herculean task. Although cotton farmers left stumps in the fields and plowed around these rotting obstructions, rice farmers had to remove all the timber in order to employ animal- or tractor-drawn machines. Wielding axes, laborers felled the trees on rice fields. To remove the standing stumps, they knelt on the ground and used crosscut saws to cut stumps off at ground level. After cutting off the stumps, farmers could use machines on the newly cleared rice fields. Because stump cutting was such back-breaking work, farmers welcomed the introduction of "stump saws" to remove these troublesome obstructions. Invented around 1910, stump saws used gasoline engines to power horizontal circular saws. Hauled from stump to stump by mule teams, stump saws enabled farmers to efficiently clear the clay-bottomed timberlands. Following the introduction of stump saws, drainage districts proliferated in the swampy White River Valley. Draining the swamplands, farmers sawed off the timber in order to create rice fields.[52]

Although rice proved a promising bottomland crop, it was not suitable for all environments in eastern Arkansas. Rice fared best on the clay-bottomed soils of the Grand Prairie and the White River Valley: The crop proved less adaptable to the silty loams of the Arkansas and St. Francis River valleys.[53] In these areas, cotton retained its hold on the land. Although the Arkansas Valley was infested by weevils in the 1910s, these pests did not overrun the St. Francis Valley until the 1920s. Given the late arrival of weevils in the St. Francis basin, this valley was the scene of a cotton boom during the early twentieth century.[54]

Weevils arrived later in northeastern Arkansas, thus this area became a refuge for Southern cotton growers. And because northeastern Arkansas possessed fertile bottomland soils, it also attracted land seekers from the Midwestern states.[55] Although there were tracts of well-drained frontlands and ridgelands in northeastern Arkansas, much of the soil was poorly drained backlands. After draining and clearing, the silty backland soils proved both durable and productive. Known as "buckshot" clay because it crumbled into tiny balls after drying, backland soils often yielded more than a bale of cotton per acre.[56] These soils were so productive that agriculturalists flocked to northeastern Arkansas, acquiring cut-over and timbered lands from corporations, speculators, and governments. Creating drainage districts in northeastern Arkansas, planters and farmers reclaimed and cleared the rich swamplands.[57] Among the more successful planters of northeastern Arkansas were two Southern-born entrepreneurs—R. E. Lee Wilson and Scott Bond. Wilson, the son of an antebellum planter, owned numerous businesses, he operated the Jonesboro, Lake City & Eastern railroad, and he amassed more than forty thousand acres in Mississippi County. In turn, Bond, a former slave, acquired a general store, three cotton gins, a gravel pit, and five thousand acres of land in St. Francis County. Joining these Southern businessmen-planters were such Northern newcomers as Governor Frank O. Lowden of Illinois, who purchased ten thousand acres near Luxora. The governor's wife, Florence Pullman Lowden, operated their plantation in Mississippi County. Among the Lowdens' neighbors was S. E. Simonson, an Illinois-born drainage engineer, who owned seven thousand acres in Mississippi County.[58] In addition to wealthy planters, there were also Southern and Midwestern farmers who acquired smaller tracts in northeastern Arkansas. Typical of such farmers were the Ohio-born brothers, Roy and Will Swank, who bought 160 acres of raw swampland in Crittenden County from a banker in nearly Marked Tree.[59] Whether they purchased small farms or acquired huge plantations, agriculturalists planted as much land as possible in cotton. In Crittenden County, for example, agriculturalists cropped more than 70 percent of the cultivated acreage in cotton, and they planted the rest of the tilled land in corn and other crops. Cotton was not only the dominant crop in northeastern Arkansas, but the plantation was also the dominant agricultural unit.[60] The "greater portion" of the cultivated land was "owned by large plantation owners and farmed on a crop-sharing basis by tenants whose chief crops" were "cotton and corn."[61]

Cotton was also grown in southeastern Missouri during the early twentieth century, but it was overshadowed by Midwestern farm crops. In Mississippi County, for example, farmers planted only one sixth of the cultivated land in cotton, and they devoted the remaining acreage to corn, wheat, and hay.[62] "The delta of Southeast Missouri" was "the last frontier of the Midwest" and the area attracted many Midwestern grain farmers who sought richer, more productive soils.[63] In southeastern Missouri, the bottomland soils yielded more corn, wheat, and hay per acre than the upland soils of Illinois.[64] Among the midwesterners who settled in southeastern Missouri was Hugh Studabaker from Bluffton, Indiana. Buying a 215-acre farm in New Madrid County from a land agency, Studabaker raised corn, wheat, cowpeas, and hogs. Losing money each year from 1909 to 1913, he finally sold his place and returned home.[65] Studabaker was one of many Midwestern settlers who failed in southeastern Missouri. Many newcomers "bought eighties [80 acres] or hundred-sixties [160 acres] and cleared them and farmed them by the labor of one family." Falling into debt, they often lost "their hard-won farms" to foreclosure. Some midwesterners, however, succeeded where others failed. Leaving Indiana in 1910, Thad Snow visited southeastern Missouri. Noting the fertile land, he decided to settle in the bottomlands of Mississippi County. Purchasing a large tract of bottomland near Charleston, he moved his family to an area where there was "pioneering enough for anybody." Although most of his land was in "deadening, cutover, and green timber," Snow did have some cleared land when he began farming. He skated on "thin financial ice" for several years, but he persevered. By 1920, he was successfully raising corn, wheat, alfalfa, hay, and cattle for market.[66] Snow did not raise cotton, but other Midwesterners learned to grow this crop. Despite the short growing season, there were no boll weevils in southeastern Missouri at this time, and cotton proved a profitable crop. Yet, the lack of labor posed a problem to cotton growers. Missourians had to import Black migrant workers to pick their cotton crops. Some of these Black workers remained in southeastern Missouri, becoming sharecroppers on local farms.[67] Between 1910 and 1920, the Black population in the six Missouri Delta counties grew from 6,301 to 7,655. The increasing Black population reflected the expanding cotton acreage in southeastern Missouri. Between 1910 and 1920, the cotton land in the Missouri Delta grew from 84,141 to 330,029 acres. The greatest cotton acreage and the largest Black population were found in Pemiscot County—a community that bordered northeastern Arkansas. By 1920, Pemiscot had 126,000 acres of cotton as well as 3,900 Black residents.[68]

Pemiscot County, Missouri, was a fascinating blend of the cotton South and the Midwestern cornbelt. Most of Pemiscot's land was in the hands of "a few men," who divided their estates "into farms of 100 to 150 acres." Landowners rented these farms "to white tenants on a cash basis," and these White renters in turn sublet land to Black croppers who worked for a share of the crop. On Pemiscot farms, the principal crops were alfalfa, corn, and cotton. Alfalfa thrived on bottomland soils, and alfalfa hay proved a valuable cash crop. In addition to alfalfa, farmers raised corn, selling the surplus or feeding it to hogs. The best land in Pemiscot, however, was reserved for cotton, and this crop was "often planted on the same land for a number of years." Despite repeated cropping, the land yielded nearly a bale of cotton each year. In order to raise cotton, Pemiscot farmers employed "the heavy [horse] teams of the corn belt," but they used "the types of implements and methods found in the cotton belt." They substituted horse power for human labor in plowing, planting, and cultivating cotton land. Despite the lavish use of horse-drawn implements, Pemiscot farmers faced a problem in harvesting their cotton, which had to be picked by hand. A shortage of cotton pickers often forced farmers to leave much of their cotton standing in the fields.[69] Although labor shortages often limited harvests, southeastern Missouri held great promise as a cotton frontier at a time when boll weevils infested the lower South.

Southeastern Missouri escaped the worst of the weevil invasion, but the Missouri Delta did not escape the major floods that periodically struck the lower Mississippi River Valley. Between 1903 and 1912, the valley enjoyed a respite from major floods. But in 1912, a record flood swept down the Mississippi River Valley.[70]

In Mississippi County, Missouri, the flood broke the levees, and floodwaters inundated the bottomlands. Families fashioned home-made boats to carry them to local ridgelands, or they sought refuge in the haylofts of barns. Although Thad Snow saved his livestock by stabling them atop an Indian mound, many farmers lost their mules, horses, cattle, and hogs. When floodwaters receded, refugee families returned to clean up their houses and put in "a right good corn crop." Following the 1912 flood, the levee district began work on a stronger riverside levee, but the embankment was incomplete when another flood struck in 1913. This flood arrived early and stayed late. Yet, floodwaters rose so slowly that families and livestock found safety on Charleston ridge. But when floodwaters retreated, farmers viewed the ruined crops and buildings that were left by the second major flood in two years.[71]

The floods of 1912–1913 also visited destruction on eastern Arkansas, northeastern Louisiana, and northwestern Mississippi. When the 1912 flood broke the levees in Mississippi County, Arkansas, floodwaters overwhelmed the town of Wilson. S. E. Simonson boated into the flooded town, where he found R. E. Lee Wilson, the wealthy businessman-planter, on the second floor of his hotel, bemoaning his fate. Wilson had lost his lumberyard, much of his Jonesboro, Lake City & Eastern railroad had been wrecked, and most of his plantation cropland had been flooded. Simonson urged Wilson to retreat to his unflooded Armorel plantation near Blytheville. At Armorel, Wilson could plant cotton on the dry land and then seed his flooded croplands as the floodwaters receded. Although R. E. Lee Wilson weathered the 1912 flood, the rebuilt levees at Wilson town broke again in April, 1913, and Wilson's financial empire suffered even greater damage during the second flood. When Simonson tried to console Wilson, he replied: "You encouraged me once before, but I am done for now." Yet, Simonson helped

Wilson secure a $400,000 loan from Midwestern creditors to "carry him through the two flood disasters."[72] The 1912–1913 floods also hit northeastern Louisiana, inundating towns and plantations. In Concordia Parish, where much of the bottomland was still unprotected by levees, the 1912 flood covered nearly all of Vidalia—the parish seat town that was located on the highest point in Concordia.[73] In contrast to the Louisiana, Arkansas, and Missouri deltas, where the levees failed, the embankments in the Y–MDLD suffered no breaks. Between 1903 and 1912, the Y–MDLD had strengthened the levee line, which withstood the record 1912–1913 floods. But in the neighboring MLD, the levee line proved more vulnerable. The 1912 flood inundated 550,000 acres in the MLD, and the 1913 flood left more than ten thousand people homeless.[74]

The costly floods of 1912–1913 left the levee districts of the Mississippi River Valley in a state of financial exhaustion. There were urgent calls for federal financial aid to improve the levee system. The Mississippi River Levee Association, which was founded in Memphis, Tennessee, after the 1913 flood, lobbied for a federal flood control bill. This lobbying effort helped to secure the passage of the flood control act of 1917. This act provided two matching federal dollars for each dollar raised by local levee districts. Although the flood control act pledged $30 million in federal aid over five years in order to rebuild the levees in the Mississippi River Valley, America's entry into World War I (1917–1918) disrupted the flow of federal aid. Only $17 million in federal money was actually spent on levee work. Fortunately, the flood control act of 1923 provided another $60 million over ten years for levee construction. Thanks to this federal largesse, the levees in the Mississippi–Yazoo basin reached a height of twenty-two feet.[75] Protected by such massive embankments, bottomland residents believed their levees were "strong enough to withstand the floods."[76]

Despite the threat of major floods and despite the boll weevil invasion, the Southern bottomlands attracted numerous investors, settlers, and workers, who were lured by dreams of cotton profits. In the years from 1900 to 1919, the demand for cotton lint and cottonseed pushed prices to spectacular levels. During these years, weevils rampaged across the lower South, reducing the supply of Southern cotton. At the same time, foreign and domestic textile mills demanded more Southern cotton. Finally, America's entry into World War I touched off a wartime economic boom, which drove up commodity prices. Thanks to wartime inflation, increased demand, and reduced supply, the farm price of cotton lint soared from nine cents a pound in 1900 to 32 cents a pound by 1919, and the farm price of cottonseed, which was crushed for oil, rose from $12.90 per ton in 1900 to $66 per ton by 1919. Because the price of cotton products outpaced production costs during this time, cotton planters often scored impressive profits. In the Mississippi Delta, for example, planters frequently received a return of more than 10 percent on their investments. Attracted by such potential profits, entrepreneurs and corporations invested heavily in bottomland acreage. The outstanding example of corporate investment was the Delta & PineLand Co. (D & PL) which created the largest cotton estate in the United States.[77]

The story of the D & PL began in 1910, when Jesse Fox, then director of the Mississippi agricultural experiment station at Stoneville, attended the world cotton conference in Brussels, Belgium. At the conference, he met the director of the Fine Cotton Spinners & Doublers Association (FCS & DA) of Manchester, England, who spoke with Fox about the possibility of raising fine-staple cotton in the Mississippi Delta. While visiting the Delta in 1911, the chairmen of the FCS & DA fell under the spell of Lant Salsbury, a Michigan land speculator, and Charles Scott, a Mississippi banker-planter. Salsbury and Scott convinced the chairmen to invest $3 million in delta lands. Purchasing lands held by Salsbury, Scott and others, the FCS & DA acquired title to thirty-eight thousand acres in Bolivar and Washington counties. To manage this empire, the British investors and their U.S. associates created a holding corporation known as the Mississippi Delta Planting Co. Because Mississippi state law forbade landholdings greater than ten thousand acres within the state, the Mississippi Delta Planting Co. purchased the charter of the moribund D & PL—a land speculation company that had been chartered in 1886 (see chapter 2). And since the D & PL had been chartered when there were no restrictions on corporate landholdings, the acquisition of the D & PL charter enabled the Mississippi Delta Planting Co. to retain its vast landholdings and yet avoid trouble with the state laws. Led by president Lant Salsbury, the D & PL operated two huge estates in the Delta. The larger estate, 33,500 acres, was located at Scott in Bolivar County, and the smaller estate, which measured 4,500 acres, was located at Estill in Washington County. In order to develop these D & PL plantations, the corporation employed a California engineer named B. J. Young. When Young arrived in the Delta, he found timbered and cut-over land that had scarcely been improved. Investing $1,500,000 in improvements, the D & PL cleared forests, drained swamplands, drilled wells, and constructed farm buildings. Development work had just begun, when the new D & PL general manager—Jesse Fox, who had left the Stoneville agricultural experiment station—arrived to battle the 1912 flood. Although floods and weevils hampered cotton production, the D & PL Co. persevered in its efforts to raise fine cotton for the U.S. market. In 1915, the corporation hired Early Ewing, the cotton breeder at the Stoneville agricultural experiment station, to develop fast-maturing fine-staple cottons. Although Ewing's breeding efforts finally resulted in the famous "Delta & PineLand" cotton varieties, the heavy investments that D & PL made in research and development reduced corporate profits. D & PL paid no dividends to its investors for nine years.[78]

Not only did D & PL investors fail to reap expected profits, but the black tenant farmers who labored on delta plantations also failed to share in the cotton money. By 1910, 92 percent of the farmers in the Mississippi Delta were landless tenants, and 95 percent of these tenant farmers were Black. Delta tenants included sharecroppers, who worked for half of the crop, as well as renters, who leased lands for cash or cotton. Black tenants preferred to rent lands, but planters preferred to employ sharecroppers. Although planters had to furnish croppers with mules, tools, seeds, and housing, the croppers had to do all the work, and they received only half of the cotton harvest as payment.[79] On many Delta cropper farms, cotton was the only crop that was grown. Croppers used mule-drawn implements to plow, bed, and plant the cotton land. After "chopping" (thinning) the cotton plants with hoes, croppers began a race with such "prevalent and troublesome weeds" as crabgrass, cockle-burs, and nutgrass. Using mule-drawn tools, croppers cultivated the cotton up to ten times during the growing season, and they hoed the crop once or twice to chop out stray weeds. When the cotton matured, the croppers and their families began the laborious task of picking the seed cotton.[80] Despite months of hard work, Delta croppers earned little money, when they settled their accounts with planters. Croppers claimed half of the cotton that they grew, but they had to repay any cash and supplies that the planters had "advanced" them during the growing season. When these expenses were deducted from their half of the cotton, croppers were lucky to receive any cash at "settling time." Most sharecroppers found it impossible to save enough money to buy their own land. Hopes of becoming landowning farmers faded as croppers' earnings failed to keep pace with escalating land prices in the early twentieth century.[81]

In Bolivar County, Mississippi, for example, it was still possible to buy land for ten to twenty dollars an acre at the turn of the twentieth century. But by 1917, land prices had risen to sixty to seventy-five dollars an acre. And during World War I, land values skyrocketed. By 1919, Bolivar land was selling for three-hundred dollars an acre. As long as cotton prices continued to rise, there seemed to be no limit to the rise in land prices.[82]

The wartime boom not only drove up land and cotton prices, but also increased the demand for livestock. Wartime demand for beef and pork encouraged bottomland planters to increase cattle and hog production. Fortunately, they had improved their livestock herds during the early twentieth century. The conquest of "tick fever" and "hog cholera" had permitted bottomland stock owners to import purebred animals to upgrade their livestock herds.

Although the scrub livestock of the Southern bottomlands had acquired a degree of immunity to tick fever and hog cholera, these endemic diseases prevented livestock improvement. Whenever stock owners tried to improve scrub cattle and hogs by introducing purebred stock, the blooded animals soon succumbed to fever or cholera. Consequently, scrub stock could not be improved until these endemic livestock diseases had been eliminated (see chapter 2).

The elimination of tick fever, nevertheless, took years to accomplish, and it required the combined efforts of the federal and state governments to eradicate this stock disease. In 1884, the U.S. Congress created the Bureau of Animal Industry to investigate endemic stock diseases.[83] Five years later, the bureau discovered that bovine fever was caused by protozoa that were transmitted by cattle ticks. When infected ticks sucked the blood of cattle, they introduced the protozoa into their bloodstream. After this discovery, the bureau began searching for a way to protect cattle from this tick-borne disease as well as a way to destroy cattle ticks. By 1895, the bureau had learned to immunize susceptible cattle against tick fever by using blood from immune scrub cattle. This discovery permitted stock raisers to immunize purebred animals in order to protect them from tick fever. The effort to eliminate cattle ticks, however, required far more research and effort. The bureau discovered that dipping cattle in vats of crude oil killed the ticks that were attached to cattle. If they repeatedly dipped cattle for several months, they managed to kill all the ticks within an area. Although crude oil killed ticks, it was expensive and difficult to transport. A chemist at Louisiana State University, however, devised an easier and cheaper method for dipping cattle. By mixing white arsenic, sodium carbonate, pine tar and water, stock raisers could prepare a cheap but potent solution that killed cattle ticks. If they dipped their cattle every two weeks from March to November, they could destroy the ticks within an area. Because cattle-dipping proved a feasible means of eliminating ticks, the Bureau of Animal Industry launched a cooperative campaign in 1906 with the Southern states in order to eradicate cattle ticks. The bureau drew a quarantine line across the lower South, and state authorities initiated cattle-dipping programs. This massive effort eventually eliminated cattle ticks within dozens of Southern counties and parishes. By 1918, the dipping programs had eradicated all the cattle ticks in the counties of northwestern Mississippi. Although cattle ticks still infested some counties in southeastern Arkansas and northeastern Louisiana, the bottomlands of northeastern Arkansas and southeastern Missouri were tick-free. After cattle ticks were eliminated from these areas, it was possible to safely introduce thousands of purebred cattle to the Southern bottomlands.[84]

The bureau not only conquered tick fever, but it also overcame hog cholera. This virulent stock disease was transmitted by microbes in the urine, manure, and blood of infected hogs. Although hog cholera was endemic among scrub hogs, which had developed a degree of immunity to this disease, cholera struck down the purebred swine that were introduced to improve scrub stock. The bureau, however, learned to inoculate susceptible hogs with serum taken from scrub hogs that had survived cholera. Thanks to cholera inoculations, it was possible to safely introduce purebred hogs to the Southern bottomlands.[85]

Although the conquest of hog cholera and tick fever allowed stock raisers to import purebred animals to improve their scrub stock, this was only the first step in livestock improvement. Stock owners also had to raise forage crops in order to feed their improved livestock. Scrub stock managed to survive on native forage in the swamplands, but improved stock needed to feed on pasture, meadow, and fodder crops in order to reach market weight.

The Southern bottomlands, however, proved an ideal environment in which to raise feed crops. Given the fertile soils, stock owners raised an abundance of meadow and fodder crops at relatively little expense. And given the long growing season, pastures gave "good grazing during eight or nine months of the year and some grazing during the other three or four months."[86]

Although known primarily for cotton, the bottomlands were well suited for raising cattle and hogs. Permanent pastures, which were sown with grasses, clovers, and vetch, sustained cattle during most of the year. To supplement permanent pastures, cattle owners raised a variety of meadow and fodder crops. Meadows, which were sown with alfalfa, lespedeza, or soybeans, yielded nutritious hay for cattle.[87] Corn and sorghum were the most important forage crops, yielding both dry and green fodder. Although corn and sorghum were often harvested when mature and fed to cattle as dry fodder, these grains had greater feeding value when harvested green. Cutting corn and sorghum plants when immature yielded tons of fresh green fodder. When stored in silos, this green fodder, or "silage," supplemented the winter pastures. In addition to these farm-grown feeds, cattle raisers also purchased cottonseed meal—a cheap byproduct of the cottonseed oil industry—which proved to be "more valuable, pound for pound, than any other feed."[88] Because swine were more omnivorous than cattle, bottomland stock owners had to raise a greater variety of pasture and forage crops in order to provide year-round feeding for their improved hogs. The best pasture crops for hogs were, "a mixture of Bermuda grass, lespedeza, bur clover, and white clover." Hogs ate the young, tender shoots, while cattle grazed on the coarser, older growth in these pastures. To supplement hog pastures, stock raisers also raised such warm-weather forage crops as soybeans, which were grazed, or "hogged-off," to save labor. After hogs grazed off the soybeans, stock owners sowed wheat, rye, and oats to provide winter grazing. Because green crops produced soft pork, it was necessary to finish hogs with corn before marketing. To save labor, swine "hogged-off" the corn fields, which were often interplanted with pumpkins and beans.[89] By raising forage, meadow, and pasture crops throughout the year, bottomland stock owners were able to raise improved hogs and cattle for regional markets.

Although livestock raising was a minor pursuit in the bottomlands before 1900, the boll weevil invasion encouraged planters to improve their livestock herds. When the weevil invasion slashed cotton yields, cotton growers sold scrub cattle and hogs to earn some money. Shipping out thousands of scrub animals, they imported purebred cattle and hogs to upgrade their herds.

The most popular purebred animals were Shorthorn, Hereford, and Aberdeen-Angus cattle as well as Duroc-Jersey hogs. The Shorthorn was the largest beef breed, but it lacked the hardiness of the Hereford and Angus breeds, which were better grazers. In turn, the Duroc-Jersey hogs were vigorous and prolific animals, which gained weight rapidly. By breeding Duroc-Jersey boars with scrub sows, and by breeding scrub cows with Angus, Hereford, or Shorthorn bulls, bottomland stock raisers rapidly improved the quality and size of their hogs and cattle.[90]

Thanks to the improvement of cattle and hogs during the early twentieth century, bottomland stock raisers captured a share of the Southern livestock market. Stock owners from the Mississippi Delta sent some meat animals to the New Orleans and Memphis markets, but they shipped "the bulk of them" to "the stockyards in St. Louis." In addition to these distant markets, the Pelty Packing Co. of Greenwood Mississippi, bought "several carloads of hogs each week," and these went back "as finished products directly to the towns and plantations of the Delta."[91] Stock raisers from southeastern Missouri also shipped animals to the St. Louis market, but stock owners in the Arkansas and Louisiana deltas sent porkers and beeves to the packing plants in Memphis, Natchez, and New Orleans.[92] Sending meat animals to a variety of regional markets, bottomland stock owners reaped the rewards from higher wartime prices for beef and pork. By 1919, live cattle fetched ten cents a pound and live hogs sold for sixteen cents per pound.[93]

By 1920, high livestock prices as well as soaring cotton prices lulled bottomland agriculturalists into a false sense of security. They assumed the prosperity of the wartime years would extend into the 1920s. But in 1920, after several years of unprecedented prosperity, the wartime economic boom collapsed. The price of agricultural commodities tumbled, and Southerners endured an agricultural recession that lasted throughout the 1920s.

4

The Southern Bottomlands During the Agricultural Recession (1920–1930)

> The Delta plantation is an economic survival from antebellum times, made possible only by an almost unique combination of soil, labor supply, and climate, which has made profitable the continued production of a specialized and valuable crop. To what extent farm machinery for cotton production could be developed to replace economically the present supply of labor, if the latter should desert the plantation, is a question that is not easy to answer with assurance. Inventors have been struggling for years with the peculiar problems involved, and cotton-picking machines are now said to be practically ready for commercial production. So far, however, hand labor is still supreme in the cotton field.[1]

After the postwar recession began in 1920, cotton prices and land values began to plummet. Between 1919 and 1926, the farm price of cotton lint dropped from thirty-two cents to thirteen cents a pound, and the price of cotton seed fell from $66 to $22.40 per ton. As cotton prices plunged, land values fell to a fraction of wartime levels. During the early 1920s, land in Bolivar County, Mississippi, sold for one hundred dollars an acre or only one third of its 1919 value.[2]

Facing a prolonged agricultural recession, planters in the Mississippi Delta responded to the crisis by planting more land in cotton. Touring the Delta in the early 1920s, Howard Snyder saw nothing but "acres and acres of cotton."[3] Although cotton prices were depressed, they were relatively higher than the prices for other farm crops. Between 1920 and 1925, Mississippi

Deltans increased cotton acreage and slashed acreage of other crops. By 1925, Deltans were planting three fourths of their tilled land in cotton.[4]

Because most of the Delta's tilled land was being planted in cotton, these expanses of cotton plants invited boll weevil infestation. To reduce weevil damage, Delta planters not only raised fast-maturing cotton, but they also applied pesticides to their cotton crops. Calcium arsenate was the only effective weevilcide (see chapter 3), but growers had to dust cotton during the night in order to poison weevils the next morning. Although airplanes dusted large areas of cotton in a short time, airplane dusting was rather costly, and only the wealthier planters used this means of weevil control. The well-capitalized D & PL Co., for example, used airplanes to dust its huge plantations.[5] Most cotton growers, however, relied on cheaper means of weevil control. Entomologists at the U.S. Department of Agriculture's Delta Laboratory at Tallulah, Louisiana, had developed affordable hand-held and mule-drawn dusters to treat cotton plants with calcium arsenate. Hand-operated dusters cost less than twenty-five dollars each, but they treated only eight acres each night. Far more effective than hand-dusters were the mule-drawn dusters. A one-mule duster cost more than $100.00, but it covered up to twenty acres in a night. A two-mule duster, in turn, sold for more than $250.00, but it treated up to thirty acres in a night. Because mule-drawn dusters treated as much acreage in a night as the size of a typical tenant farm (twenty to thirty acres), they proved ideal for Delta plantations. Yet, mule-drawn dusters could be used only on fields without stumps. On "new ground" fields, which were dotted with stumps, it was impossible to use efficient mule-drawn dusters, and it was necessary to use the inefficient hand-dusters to poison weevils. The need to use mule-drawn dusters prompted many Delta planters to remove the stumps from new cotton fields. They paid workers about twenty-five cents per stump to grub up and remove these obstructions.[6]

The removal of stumps from fields not only permitted the use of mule-drawn dusters, but it also allowed planters to use tractors for tilling cotton lands. By 1920, U.S. manufacturers had developed effective gasoline-powered tractors for farm use. The most popular of these tractors, Henry Ford's "Fordson," proved useful for plowing land, but it was too cumbersome to cultivate row crops such as cotton. But by 1924, the International Harvester Company was marketing the "Farmall"—a tricycle-wheeled tractor whose two closely spaced front wheels were designed to travel between crop rows.[7] Although the Farmall could be used for cultivating cotton as well as plowing land, it did not immediately replace the mule in the Mississippi Delta. Tractors cost hundreds of dollars, consumed costly fuel, were prone to mechanical breakdowns, and had a use life of less than eight years. In contrast, a good mule cost about one hundred dollars, ate cheap fodder, was remarkably healthy, and had a work life of about eight years. Tractors were useful for plowing cotton fields, but sure-footed mules, when directed by experienced plowhands, proved unsurpassed in cultivating cotton. And because there were no cotton-picking machines to attach to tractors, planters were unable to replace the tenant families who picked the cotton crops.[8] Consequently, tractors plowed cotton lands, mules cultivated cotton plants, and tenant families picked seed cotton on Delta plantations.

Because Delta planters depended on tenants to pick and cultivate cotton, they tightened their hold on these landless people. Black tenants had always preferred renting land to sharecropping. Renters took the risk that they could lose everything if the crop failed, but they stood to gain if the crop prospered. Renters had more acres to work, they raised more provisions for themselves, and they had a higher income than sharecroppers. They also had more success in resisting the planters' efforts to supervise their work.[9] Planters, however, preferred to hire sharecroppers rather than to rent farms. Planters claimed half of the sharecroppers' cotton, and they closely supervised the cropper's efforts. Croppers also raised fewer provisions on their tiny plots, and they had to buy most of their supplies from the planters. Earning more money from croppers, planters tried to demote renters to sharecroppers. Sharecropping, however, reduced Black families to penury. Given their meager incomes, most delta sharecroppers lived in flimsy shacks, they dressed in shabby clothes, and they survived on a debilitating diet of little more than cornmeal, flour, and salt pork.[10]

Although Black tenants were sinking deeper into poverty, Delta planters were expanding their cotton acreage by reclaiming more swampland. By 1923, the Mississippi Delta contained forty-four drainage districts, which were reclaiming 2,035,000 acres or almost half of the Delta's total area. The bulk of this reclaimed bottomland was "buckshot" clay soil, which was remarkably fertile. Eager to put this reclaimed soil into cultivation, planters contracted out the tedious work of land clearing. Gangs of Black workers slashed the forest growth, burned the slash, reburned the charred timbers, and often grubbed up the stumps. After such through clearing, the drained "buckshot" soil yielded over a bale of cotton per acre without the use of any fertilizer.[11]

Cultivating the richest cotton lands in the South, Delta planters were determined to secure the best prices for their fine-staple cotton. Meeting in Memphis, Tennessee, cotton planters formed the Staple Cotton Cooperative Association (SCCA) in 1921. The SCCA, which established headquarters at Greenwood, Mississippi, was a planter-controlled marketing association that served the Delta counties. The SCCA stored members' cotton in safe warehouses until prices were most favorable. Marketing cotton when prices peaked, the SCCA secured higher profits for members. The marketing program proved so successful that the SCCA also began financing cotton crops in 1924. The Association Discount Corporation loaned funds to members at lower rates than commercial lenders. Finally, the SCCA even offered flood insurance to bottomland planters who lived under the threat of major floods.[12]

Thanks to cooperative marketing as well as cost-cutting measures, many Mississippi Delta planters survived the agricultural recession of the 1920s. Yet, few prospered during this time. Writing in 1930, agronomists at the Delta agricultural experiment station in Stoneville admitted that "few Delta plantations have operated at a profit for the past 10 years," and they noted that there was "an alarming increase in the mortgaged indebtedness" of plantations.[13]

Despite diminished profits, Delta planters continued to raise cotton, and they continued to supply Memphis, the world's largest inland cotton market, with cotton products. Although riverboats still carried some cotton products to market, the "major portion" of the cotton bales traveled by rail to Memphis. But by the late 1920s, there was "an increasing tendency to truck the bales" directly to the Memphis market.[14]

During the early twentieth century, railroads drove most of the steamboats from the Mississippi River Valley. Steamboats charged less than railroads to ship commodities to market, but steamers were slower and less reliable than railways. By 1925, fewer than forty aging steamboats operated on the Mississippi River. In addition to these surviving Mississippi riverboats, a small fleet of steamers and gasoline-powered launches worked the Yazoo River, delivering Delta products to the Memphis market. Although Delta planters rarely consigned valuable cotton bales to sluggish riverboats, they still used boats to ship sacks of cotton-seed to the Memphis oil mills.[15]

In contrast to the slow steamers that hauled cottonseed, powerful locomotives whisked the Delta's cotton bales to Memphis and other markets. Between 1900 and 1920, the Delta's railway trackage grew from 718 to 1,019 miles.[16] Yet, virtually all of this trackage was in the hands of two railroad corporations. The Yazoo & Mississippi Valley railway, a subsidiary of the Illinois Central system, controlled four north–south routes through the Mississippi Delta. In turn, the Southern Railway owned the east–west route across the Delta until 1923. In that year, the newly incorporated Columbus & Greenville (C&G) railroad acquired the Southern Railway's east–west route. Blanketing the Delta, these railroads offered convenient service to regional markets, but they monopolized transportation in the Mississippi Delta.[17] The only challenge to this rail monopoly came from the growing public highway system.

At the turn of the twentieth century, the Delta's public roads were little more than "muddy impassable trails, winding through the dense forests and impenetrable cane brakes, connecting the more frequent points of habitation."[18] On such muddy roads, it cost as much as twenty-three cents per ton-mile to haul cotton bales in mule-drawn wagons. Given such high hauling costs, dirt roads served as little more than feeders for railroads.

Growers hauled cotton over dirt roads at great cost to railroad depots, where bales were then shipped on railways for little more than one cent per ton-mile. In order to lower the high cost of road-hauling, it was necessary to grade, drain, and surface the public roads. In 1900, the Mississippi state legislature passed a road contract law, permitting counties to make contracts to improve and maintain public roads. If counties improved their public roads by adding rolled gravel surfaces, "all-weather" roadways were created, which allowed growers to haul cotton in wagons more quickly and reliably to railroad stations. The cost of hauling on rolled gravel roads fell to thirteen cents per ton-mile. Despite the advantages of gravel roads, such improved roadways were rare before World War I. By 1910, only 125 miles of Mississippi's public roads had gravel surfaces.[19] Even these gravel roads were inadequate for the new motor trucks that were rolling out of U.S. factories. Trucks required asphalt-surfaced or concrete-paved roads for most efficient operation. Yet, asphalt and concrete roads cost three or four times as much to construct as gravel roads. In order to build such expensive roads, counties had to raise construction funds by issuing road bonds, taxing motor vehicles, and charging license fees. Despite these new revenues, Mississippi counties were unable to improve most public roads. But in 1916, the U.S. Congress passed a federal highway act to aid road-building. This act provided matching federal funds to state highway departments for rural road construction. Creating a state highway department in 1916, Mississippi began building federal-aid highways. In 1921, a new federal highway act provided federal matching funds for intercounty and interstate highway construction. By 1922, Mississippi claimed 3,357 miles of federal-aid highways, 5,400 miles of state roadways, and 44,328 miles of county public roads. Unfortunately, the bulk of these roadways were still dirt roads. In order to raise more money for highway improvement, the state of Mississippi imposed a one-cent per gallon gasoline tax in 1922, The state highway department received 40 percent of this tax, and the rest went to the county governments. By 1924, the gasoline tax was raised to three cents a gallon, and the proceeds were divided equally between state and county governments.[20] Thanks to improved funding, the Mississippi Delta counties redoubled their road-building efforts, leading the state in highway construction. Bolivar County, for example, claimed more miles of improved highway than any county in the state. Although most of the Delta's improved roads were surfaced with gravel, the finest highways were paved with asphalt or concrete. Paved roads withstood wagon, auto, and truck traffic far better than gravel roads, providing hard, smooth surfaces that dramatically cut hauling costs. It cost only one cent per ton-mile to haul cotton bales by motor trucks on paved concrete highways.[21] Because the cost of shipping on paved highways approached the cost of shipping on railroads, improved roadways emerged as competitors to railroads in the Mississippi Delta.

Although county, state, and federal governments invested thousands of dollars in Delta highways, this investment paled beside the millions that the levee districts and the federal government spent on the Delta levee line. By 1927, the Y–MDLD, which covered the northern Delta counties had completed ninety miles of levee embankments along the Mississippi River front. Because the Y–MDLD levees were considered complete by 1917, only local work was needed to bring the levee line up to the standards set by the Mississippi River Commission. The commission recommended levees that were 25 feet high and 212 feet wide at the base. Such massive embankments were believed high enough to repel floodwaters and wide enough to prevent undermining by burrowing crayfish. In contrast to the imposing levees of the Y–MDLD, the levees of the MLD, which covered the southern delta counties, needed massive work to bring them up to commission standards. The federal government spent more than $10 million and the MLD spent more than $8 million on levee work in the decade after 1917. Although the levees finally reached commission grade by 1927, there was a worrisome gap in the levee line. The MLD had completed 190 miles of levee line along the Mississippi River front, but the 30 mile gap between Brunswick and Vicksburg was still under construction by 1927. Despite the Brunswick gap, it was widely believed that the MLD and Y–MDLD levees would hold back the highest floods. Yet, the building of levees in northeastern Louisiana, eastern Arkansas, and southeastern Missouri posed a threat to the Mississippi Delta because these westside levees confined floodwaters to the swollen Mississippi River channel.[22]

In northeastern Louisiana, the levees of the Fifth Louisiana and Tensas Basin levee districts (see chapter 3) protected the railways, highways, and plantations of the delta counties. Northeastern Louisiana not only contained the venerable V,S&P railroad, but the Delta also held the New Orleans & Northwestern (NO&NW) line, which ran from Vidalia on the Mississippi River to Rayville on the V,S&P, as well as the new Memphis, Helena & Louisiana (M,H&L) line, which ran from Ferriday, Louisiana, to Helena, Arkansas, and Memphis, Tennessee. Both the NO&NW and the M,H&L lines were part of the huge Missouri Pacific system that controlled routes from St. Louis to Memphis to New Orleans. Providing access to these regional cotton markets, the Missouri Pacific lines in northeastern Louisiana were supplemented by a network of parish and state roads. By the 1920s, gravel roads snaked across the bottomlands of the Louisiana Delta. Local roadways and regional railways served the needs of the Delta's plantations, which still suffered from boll weevil infestations. But by the 1920s, Delta cotton planters were coping with the presence of weevils. New fast-maturing cotton varieties replaced the old slow-growing types that were overrun by weevils. To further reduce weevil damage, planters applied nitrogenous fertilizers to hasten the growth of plants, and they dusted cotton with calcium arsenate poison. Thanks to such innovations, cotton production recovered somewhat during the 1920s. Adapting to the troublesome boll weevils, Louisiana's cotton planters felt secure behind their levee embankments.[23]

In neighboring Arkansas, a patchwork of levee districts protected the Delta counties (see chapter 3). During the 1920 flood, the Delta levees held back floodwaters that lapped five feet below the tops of the embankments. Following this flood, engineers believed that the Arkansas bottomlands were "protected forever" from "Mississippi River Floods."[24]

Protected by the Delta levees, railroad companies in eastern Arkansas pursued a strategy of construction and consolidation during the early twentieth century. Among the newly constructed lines were the Missouri & North Arkansas (M&NA) railway and the Jonesboro, Lake City & Eastern (J, LC&E) railroad. The M&NA, which was nicknamed the "May Never Arrive," began its career as the St. Louis & North Arkansas (St.L&NA) in 1899. Reorganized as the Missouri & North Arkansas in 1906, the company planned to extend the line across eastern Arkansas. Completing construction by 1909, the M&NA stretched from Joplin, Missouri, to Helena, Arkansas. Hauling lumber and cotton, the M&NA had to compete with major lines for the east Arkansas trade, and there was not enough freight to support such a lengthy line. Running deficits during most years, a group of St. Louis capitalists bought the M&NA, but the railway continued to struggle. A more successful line, at least for a time, was the Jonesboro, Lake City & Eastern railroad, which penetrated the "sunk lands" of the St. Francis River Valley. The J,LC&E line was chartered in 1893 by a group of Jonesboro businessmen, who planned to exploit the timber of northeastern Arkansas. Building to Lake City in 1899, the J,LC&E bridged the St. Francis River and reached Blytheville by 1903. Proving prosperous, the JLC&E attracted the attention of the wealthy businessman-planter, R. E. Lee Wilson of Mississippi County. Wilson purchased controlling interest of the railway in 1911, and he added his own Wilson & Northern (W&N) line to the company. Reaching a length of ninety-six miles, the J,LC&E suffered much damage during the 1912–1913 floods. Additionally, the timber business began to slacken by the 1910s. Yet, Wilson responded to these challenges by encouraging farmers to settle on the cut-over lands that bordered the J,LC&E line. Wilson's land agents sold tracts to farmers who planned to raise cotton on reclaimed swamplands. As cotton prices soared during the 1910s, these farmers raised thousands of cotton bales, which the railway shipped to market. This new-found prosperity, however, faded after 1920, when cotton prices plunged. By 1925, it was necessary to sell the J,LC&E to the St. Louis and San Francisco (Frisco) railroad.[25] The Frisco company had entered eastern Arkansas in 1901, when it leased the Kansas City, Ft.Scott & Memphis line (see chapter 3). The following year, the Frisco line acquired Louis Houck's rail system, which stretched from southeastern Missouri into northeastern Arkansas (see chapter 2). Pushing into eastern Arkansas, the Frisco line had to compete with the Rock Island, the Cotton Belt, and the Missouri Pacific railways for the local traffic. The Chicago, Rock Island & Pacific (Rock Island) company entered the Arkansas Delta in 1904, when it leased the Choctaw, Oklahoma & Gulf line for 999 years. To generate freight for this newly leased line, the Rock Island company encouraged rice growing on the Arkansas prairies. The Rock Island line, however, had to compete with the established St. Louis–Southwestern (Cotton Belt) railway for the rice and cotton freight of eastern Arkansas. Although the Cotton Belt, the Rock Island, and the Frisco lines controlled much of the trackage in the Arkansas Delta, these companies were overshadowed by the Missouri Pacific—the largest railway system in Arkansas. The Missouri Pacific controlled the old Iron Mountain route as well as two main lines that terminated in Memphis.[26] Offering rail service to the Memphis, St. Louis, and New Orleans markets, the lines of the Missouri Pacific, the Cotton Belt, the Rock Island, the Frisco, and other companies blanketed the bottomlands of eastern Arkansas.

These railway companies drove most of the competing riverboats from the waterways of eastern Arkansas during the early twentieth century. By the 1920s, there was little traffic left on the rivers, except for "a few steamers" that plied the Mississippi River.[27] Although railroads dominated transportation in eastern Arkansas, this rail monopoly was challenged by improved highways and motor trucks.

Highway improvement in Arkansas began in 1899, when the state legislature permitted the creation of road districts within counties. Local districts built graded dirt roads that served as feeders for the graveled state highways. Some of these graveled state roads in turn became part of the interstate highways that crossed the Arkansas Delta. The Mississippi River Scenic Highway, the Bankhead Highway, the Ozark Trail, the International Trail, and the King's Highway radiated out from Memphis, Tennessee, which became a highway as well as a railway center. Because these improved highways linked the Arkansas Delta to Memphis, roadways offered cotton growers an alternative to railways. Hard-surfaced roads permitted motor trucks to haul cotton bales directly to the Memphis market. On one fall day in 1923, a geographer counted three hundred motor trucks, most of them laden with cotton bales, driving southward along the paved King's Highway to Memphis. By the 1920s, motor trucks handled much of the agricultural freight in eastern Arkansas, and trucks helped drive such smaller railways as the Jonesboro, Lake City & Eastern to financial ruin.[28]

Although Arkansans enjoyed convenient road and rail transportation, they endured poor commodity prices during the 1920s. The farm price of cotton lint hovered around nineteen cents a pound, and the price of cotton seed averaged about thirty-one dollars in this decade.[29] Despite depressed prices, Arkansans continued raising cotton during the agricultural recession. Hoping to win the best price for their cotton, planters formed a marketing cooperative—the Arkansas Cotton Growers' Association (ACGA)—to withhold cotton from the market until prices were most favorable.[30] Many growers, however, were unable to wait until the cotton market improved. They had to sell cotton as soon as it was harvested in order to pay their debts. Farm "debt was tremendous," and planters tried to raise more cotton "in order to pay their creditors."[31]

To worsen matters, boll weevils invaded northeastern Arkansas.[32] Roy Swank, a Crittenden County farmer, recalled that weevils arrived around 1921. After many months of hard labor, Roy and his brother Will had cleared 30 acres of their 160-acre bottomland farm. In 1921, they hired a White sharecropper who planted six acres in cotton and five acres in corn. After the weevils arrived, the cropper harvested only three bales from his cotton patch. The next year, the Swanks rented land to two White families from Hattiesburg, Mississippi. Renting land on shares, the Mississippians promised to pay the Swanks one third of the corn and one fourth of the cotton that they raised. These families planted more than forty acres in cotton and more than ten in corn. Boll weevils, however, swept across their fields, leaving them with only ten bales of cotton. Discouraged by this poor harvest, the families returned to Mississippi. Ironically, they had left Mississippi in order to avoid weevils, but these pests had moved northward. A neighbor of the Swanks, a Mississippian named "Big" Cooper, had also moved north to escape the boll weevil. He bought an estate on the St. Francis River, which he named "Boll Weevil Plantation." After weevil overran his fields, Cooper sat "on his front porch beside the river," and he "twanged out sad blues songs on his battered old guitar."[33]

Boll weevils, of course, posed no threat to the Arkansas rice crops, but low prices stifled the state's once thriving rice industry. In 1920, when rice was selling for $1.61 per bushel, Arkansans sowed 175,000 acres of rice. The following year, when rice fell to one dollar a bushel, they sowed fifty thousand fewer acres.[34] Because prices stagnated around one dollar a bushel, rice farmers formed the Arkansas Rice Growers Cooperative Association (ARGCA) to weather the hard times. The ARGCA extended credit to beleaguered growers, and even acquired the Stuttgart Milling Company in order to process members' rice. Although rice farmers survived the recession better than cotton growers, low prices forced the abandonment of much rice land.[35] Unable to pay property and drainage taxes on their acreage, some Arkansas rice farmers even lost tax-forfeited lands to the state.[36]

While Arkansans struggled with low prices and crop pests, southeast Missourians enjoyed a measure of prosperity during the 1920s. Despite depressed cotton prices, southeast Missouri escaped the worst of the boll weevil invasion, and the area dramatically expanded cotton acreage. Protected by the levees of the St. Francois Levee District (see chapter 3), southeast Missourians raised cotton and grains on alluvial soils, and they shipped these commodities to market on railways and roadways.

In southeastern Missouri, three major railway systems—the Frisco, the Cotton Belt, and the Missouri Pacific—controlled the north–south routes that ran between St. Louis and Memphis. Radiating from these trunk lines were numerous branch lines that provided local service for bottomland farmers. In addition to this dense railway network, southeastern Missouri contained several gravel and concrete highways that followed the ridgelands through the bottoms. Such improved roads as the King's Highway, which led to Memphis, offered reliable routes for motor trucks. Although trucks economically hauled such valuable commodities as cotton bales to market, only railways efficiently shipped such bulky, low-value commodities as wheat and corn.[37]

Thanks to efficient rail service, most Missouri Deltans had specialized in grain farming during the early twentieth century. Raising wheat on the ridgelands and growing corn on the bottomlands, farmers harvested bountiful grain crops, which they shipped by rail to the St. Louis market. Grain elevators and flour mills lined the railways that ran through southeastern Missouri. Grain farming boomed during World War I, when wartime demand pushed prices to new heights. But in 1920, grain prices began to tumble. Due to falling prices, wheat growing was no longer profitable. Because it cost $18.97 in 1922 to raise an acre of wheat, and because an acre of wheat grossed only $14.35, growers lost $4.62 on each acre. Corn, in turn, yielded only a meager profit. It cost $23.60 to raise an acre of corn in the Missouri Delta and it grossed only $30.50 per acre.[38]

Although grain farming yielded little profit, cotton offered greater rewards. In 1922, an acre of Missouri Delta cotton yielded a value of $60.40. It cost $33.57 to produce an acre of cotton, thus the crop earned a net value of $26.83 per acre. Proving far more profitable than corn and wheat, cotton offered hope to despairing grain farmers in the Missouri Delta.[39]

Cotton long had been grown in the southernmost counties of the Missouri Delta (see chapters 2 and 3). But after 1920, cotton farming spread across southeastern Missouri. Although the cold winters of southeastern Missouri checked the advancing boll weevils that arrived in the 1920s, the untimely frosts hindered cotton growing. The cotton growers adapted to the short growing season by raising fast-fruiting cotton varieties that matured before fall frosts, and they applied fertilizers to cotton in order to hasten maturity. Adapting to the colder climate, cotton farmers even formed a marketing cooperative—the Missouri Cotton Growers Association (MCGA)—to gin, store, and sell the crops from southeastern Missouri.[40]

Successfully raising cotton in most of the Missouri Delta, farmers faced labor shortages because cotton had to be harvested by hand. Local farmers, who wanted to raise cotton, had to hire Black migrant workers to pick their crops, and Southern planters, who moved to southeastern Missouri in order to escape the boll weevils, had to import Black tenant families to plant, cultivate, and harvest their cotton. The arrival of these Black workers and tenants altered the racial composition of southeastern Missouri. Between 1920 and 1930, the Black population of the six Missouri Delta counties grew from 7,655 to 23,338 residents.[41]

Planters and farmers welcomed the Black newcomers, and they "treated them with cordiality, because they could grow cotton" on "unprofitable corn and wheat land."[42] Yet, the influx of Black Southerners angered the local landless Whites, who competed with the Black newcomers for jobs. Periodic outbursts of racial violence punctuated the 1920s. Despite this random violence, most Black newcomers found jobs as tenants and laborers on Missouri Delta plantations and farms.[43]

Thad Snow, a Mississippi County farmer, remembered the arrival of the Black croppers and renters during the early 1920s. To house these new arrivals, Snow and other landowners "threw up thousands of shaky 'box' houses" on their farms and plantations. Tenants not only had to be housed, but they also had to be furnished with food and clothing while they raised the cotton crop. Because furnishing tenants with supplies and housing cost extra money, Missouri cotton growers turned to Memphis cotton merchants for credit. Borrowing money from Memphis creditors, Missouri planters and farmers raised cotton with Black tenants and workers. Although growers "had some bad luck, bad seasons and low prices" in the first years, cotton usually improved their economic circumstances. Snow claimed that cotton "gave us more credit, prolonged our financial decline, and somewhat delayed our final debacle."[44]

Cotton not only saved the finances of southeast Missourians for a time, but it also transformed the landscape. Before 1920, the Missouri Delta was essentially an outpost of Midwestern grain agriculture. It was largely a land of grain fields, family farms, grain elevators, and flour mills. After 1920, southeastern Missouri became an outlier of southern cotton agriculture. Cotton plantations, tenant farms, cotton gins, and cottonseed-oil mills now filled the landscape. By 1927, southeastern Missouri more closely resembled the Southern cotton belt than the Midwestern cornbelt.[45]

Although southeastern Missourians survived the agricultural recession by shifting from grain to cotton, cotton growers in eastern Arkansas, northeastern Louisiana, and northwestern Mississippi sank deeper into debt. They prayed for an upturn in crop prices. After seven years of agricultural recession, the inflated prices of the World War I era seemed like a distant memory.

World War I had not only driven crop prices to stratospheric heights, but had also directed national attention to malaria—a dangerous disease that still haunted the Southern bottomlands. During the early twentieth century, eastern Arkansas, northeastern Louisiana, northwestern Mississippi, and southeastern Missouri were regarded as hotbeds of malaria. Because the only effective treatment for malaria was quinine, it was said that Arkansas swamplanders begged for quinine capsules like spoiled children cried "for striped candy."[46] In the Arkansas swamps, the "little frogs" piped "in their shrill voices, 'Quinine,' 'quinine,' 'quinine,' while the big bullfrog with his deep bass voice" chimed "in 'Double the dose'!"[47]

Although quinine eased the symptoms of malaria, public health officials believed that the best way of combating this disease was to destroy the malaria-bearing mosquitoes. In 1916, the U.S. Public Health Service, the Arkansas Board of Health, and the Rockefeller Foundation conducted a mosquito control experiment in Crossett, Arkansas. Squads of health workers roamed through Crossett, making war on mosquito larvae in pools, ponds, and streams. They cut ditches to drain stagnant waters that harbored larvae, applied fuel oil to ponds in order to kill larval mosquitoes, and even introduced minnows to gobble up larvae in sluggish streams. Thanks to these efforts, mosquito populations plummeted in Crossett, and the town's malarial cases dropped dramatically.[48]

Despite such successful examples of mosquito control, the public campaign against malaria stalled in the Southern bottomlands during the 1920s. Given the long agricultural recession, county and state governments lacked the resources needed to eradicate anopheline mosquitoes. There were, however, some local successes in the public fight against malarial mosquitoes. Railroad companies often furnished fuel oil to the towns along their routes, thus allowing townspeople to treat stagnant waters. In addition, airplanes sprayed fuel oil and "Paris green" over the Arkansas rice fields, where mosquitoes bred. The greatest victory in the public war against malarial mosquitoes, however, was achieved by the local drainage districts.[49] Drainage projects had reclaimed thousands of acres of swampland, where anopheline mosquitoes had once flourished (see chapter 3). As drainage work progressed, malarial swamplands were transformed into salubrious farmlands.[50]

The draining and clearing of thousands of acres of swampland became the second greatest achievement in the settlement of the Southern bottomlands. The greatest achievement was the building of a huge levee line that protected these reclaimed acres. Following the 1912–1913 floods, the rebuilt levee line withstood several inundations. By 1927, the levees were believed strong enough to repel the highest floodwaters. Yet, no one reckoned with the 1927 flood—the most destructive in the history of the Southern bottomlands.

The flood of 1927 began in the fall and winter months of 1926, when heavy rains pelted the central United States, swelling the upper tributaries of the Mississippi River. On January 1, 1927, the Mississippi River reached flood stage at Cairo, Illinois—the earliest for any year on record. As the rains continued, the lower tributaries of the Mississippi began to flood. By April 15, the Mississippi river was in flood stage from Cairo to the Gulf of Mexico.[51]

The rising Mississippi River threatened the levee line that protected southeastern Missouri, eastern Arkansas, northwestern Mississippi, and northeastern Louisiana. On April 16, the levees broke at Dorena, Missouri, thirty miles below Cairo, and floodwaters inundated nearly two hundred thousand acres in southeastern Missouri. Three days later, the levees burst at New Madrid, Missouri, flooding one million acres in southeastern Missouri and northeastern Arkansas. Yet, these breaks in the upper level levee line did not relieve the pressures building on the levees that guarded the Mississippi Delta.[52]

Levee districts in the Mississippi Delta mobilized local men for the high-water fight. Working twenty-four hours a day, they tried to shore up the threatened levees with rows of sandbags. Despite heroic efforts to strengthen the levees, the Mississippi River kept rising. Hoping the levees would hold, Delta planters tried to put in a spring crop of cotton. After several years of poor prices, many planters were "heavily mortgaged," and their tenants were also mired in debt. Mortgaged planters were reluctant to evacuate the threatened Delta. They not only feared losing their cotton crops, but they also feared that their indebted tenants would slip away.[53]

The rising river, however, burst the levee at Mounds Landing on the morning of April 21. A wall of floodwater poured through the widening break. Deltans fled for their lives, abandoning homes, furnishings, tools, and livestock to the rampaging waters. Within ten days, the flood covered one million Delta acres under ten feet of muddy water. On the same day that the levees broke at Mounds Landing, the Mississippi River burst the Arkansas levee line at Medford and Pendleton, and floodwaters surged across southeastern Arkansas.[54]

Responding to the flood crisis, the Red Cross began setting up temporary camps for flood refugees. Yet, the widening flood crisis proved too much for the limited resources of the Red Cross. On April 22, the day after the Mississippi and Arkansas levees broke, President Calvin Coolidge appointed Secretary of Commerce, Herbert Hoover, as chairman of a select committee, which included the secretaries of the war, navy, treasury, and agriculture departments. With the addition of the Central Committee of the Red Cross, it became the Mississippi Flood Committee. Mobilizing federal and private resources, the committee launched a campaign to rescue flood victims, aid flood refugees, and rehabilitate the flooded areas.[55]

As the committee's rescue workers were saving flood victims in Mississippi, Arkansas, and Missouri, the Mississippi River was threatening the levees of northeastern Louisiana. Local White and Black men, even high school boys, shored up the failing levees. Behind the levees, planters were putting in their cotton crops. It "was a gamble for the planter; if the flood came, seed, labor, wages—all would be lost; if not, why, he was that much ahead of the game." But when waters kept rising, planters began moving their mules to higher ground.[56] On May 2, the swollen river burst the Louisiana levee line at Cabin Teele, and floodwaters poured over northeastern Louisiana, creating an inland sea that stretched from Vicksburg to Monroe.[57]

When floodwaters finally flowed southward to the Gulf of Mexico, some farmers in southeastern Missouri and eastern Arkansas straggled back to their fields, hoping to plant cotton on the muck-covered land. Floodwaters also receded in the Mississippi Delta, exposing some tracts of higher ground that were also planted in cotton. But as the cotton seeds were sprouting, another flood began moving southward from Cairo. This "June rise" poured through the breaks in the levee line, flooding the bottomlands once again. As late as July, over one million acres were still under water. It was not until late August that the last of the floodwaters receded from the bottomlands.[58]

Before receding into history, the 1927 flood inundated most of southeastern Missouri, eastern Arkansas, and northeastern Louisiana. In southeastern Missouri, New Madrid County, where floodwaters covered 193,000 acres, suffered the worst flood damage. In turn, Desha County, where floodwaters inundated 478,000 acres, endured the greatest damage in eastern Arkansas. Finally, Concordia Parish suffered the worst flood damage of any parish in northeastern Louisiana. Floodwaters covered 457,000 acres in Concordia, and the town of Vidalia, which sat on the highest point of the parish, was under two to five feet of water. Although the lower Mississippi Delta was also deluged by the 1927 flood, the upper Delta escaped the worst of the flooding. The well-built levees of the Y–MDLD withstood the floodwaters, and the northern Delta counties suffered only minor damage. A geographer, who drove through the upper Delta during July 1927, found undamaged plantations and growing cotton fields. But as he neared the Mounds Landing break, he saw the devastated lower Delta. Within the lower Delta, Washington County, where floodwaters inundated 458,000 acres, suffered the worst damage. The flood destroyed 2,150 houses and damaged another 13,250 houses and stores. In addition, floodwaters drowned 1,500 mules and horses, 3,500 cattle, and 10,000 hogs. Finally, the deluge ruined $11 million worth of seeded cropland in Washington County.[59]

Washington and neighboring Bolivar County contained the most heavily damaged plantation in the Mississippi Delta. The huge D&PL estate lay in the path of the floodwaters that surged through the Mounds Landing break. The estate lost 1,250 acres of farm land to erosion, sand deposits covered another 3,500 acres, and floodwaters ruined the seeded crops on the remaining cultivated land. The flood also drowned two hundred mules, and it swept away more than one hundred plantation buildings. The total flood damage to the D & PL estate was estimated at five hundred thousand dollars.[60]

Although floodwaters inundated the D & PL estate, Oscar Johnston, the newly appointed D & PL manager, rescued the Black tenant families from the flooded plantations, assembling them at the Deeson refugee camp in the unflooded upper Delta. Receiving relief supplies from the Red Cross, Johnston's foremen distributed these goods to his encamped tenants. To prevent any D & PL tenants from escaping, armed National Guardsmen patrolled the Deeson camp perimeters.[61]

The experiences of the D & PL tenants in the Deeson refugee camp were not unique. Black refugees in other Red Cross camps often received shabby treatment. Refugees "in hundreds of cases were forced to work at the point of guns on the levees long after it was certain that the levees would break," and they received only paltry wages for their "hard and dangerous work in fighting the flood." In addition, the officials in charge of flood relief, "whether unwittingly or not," permitted the relief organizations to be used by planters in order to retain control over refugee tenants. Red Cross officials often issued relief supplies in bulk to planters, who in turn distributed these goods to refugee tenants. Planters frequently charged tenants for these relief supplies, adding the cost of these goods to tenants' debts. Because planters feared that Black tenants might escape from refugee camps, armed militiamen patrolled the encampments. Black refugees were released only to planters who came to the camps, identified their tenants, and then took "these Negroes [*sic*] back to the plantation from whence they came." Despite the armed guards, some Black refugees escaped "to shift for themselves." They chose "to forego food, shelter, clothing and medical attention rather than go back to virtual slavery on the plantations from which the flood waters had driven them." Most refugees, however, remained in the armed camps, subsisting on Red Cross rations until they were released to their planters.[62]

Disheartened by their experiences in the Red Cross refugee camps, tenants returned to their old plantations when floodwaters receded. Before leaving the camps, the refugees received sacks of cotton seed and packets of vegetable seeds from Red Cross workers. Because floodwaters receded too late to plant the cottonseed during the summer of 1927, tenants scratched in the garden seeds.[63] A late summer drought, however, withered most of the vegetables, except the reliable turnips. As one journalist exclaimed: "Thank God for turnips!"[64] Subsisting on vegetables and "a small food allowance" from the Red Cross, tenants survived the waning months of 1927. After living for months on meager rations, many tenants were showing signs of pellagra—a debilitating disease that was caused by too little niacin in the diet. Although the Red Cross saved pellagra victims by giving them brewer's yeast, many tenants saw little future in staying on ruined plantations.[65]

Because bottomland levees, towns, railways, and highways needed rebuilding after the flood, tenants found wage work on reconstruction projects. Saving some money, they bought train tickets to the Northern cities. Thousands of disgruntled Black tenants fled the Southern bottomlands after the 1927 flood.[66]

The great flood not only drove away many Black tenants, but it also dimmed the hopes of many White planters. After years of poor crop prices, many indebted planters lost their 1927 cotton crop to the flood. Because floodwaters receded too late to plant cotton, they tried planting soybeans, using seeds that were furnished by the Red Cross. Although soybeans sprouted on the muddy land, late summer droughts and early autumn frosts doomed this desperation crop.[67] When the spring of 1928 arrived, planters were desperate to plant cotton. They needed credit to repair their plantations and to furnish their tenants with provisions. Because many banks in the flooded districts were in "bad shape," these institutions were often unable to loan funds to local planters. Fortunately, the Mississippi Flood Committee organized agricultural finance corporations in Mississippi, Arkansas, and Louisiana in order to make low-interest loans to flood-stricken planters. In addition to these agricultural reconstruction loans, the American Red Cross furnished cash grants and relief supplies to local Red Cross committees in the hardest hit counties. These committees channeled much of this aid to local planters. Thanks to this timely aid, many bottomland planters managed to put in crops during the 1928 season.[68]

Dispensing aid to planters and tenants after the great flood, the Red Cross hoped to restore the status quo in the southern bottomlands. Yet, Red Cross flood relief literally and figuratively planted the seeds of change. Tenants received vegetable seeds that improved their diets, and planters in turn received legume seeds that diversified bottomland agriculture.

Although the vegetable seeds that Red Cross workers distributed in 1927 were blasted by drought, Red Cross committees issued more than forty-nine thousand garden seed packages in 1928 to flood-stricken communities. A typical Red Cross seed package included, tomatoes, beets, beans, peas, corn, okra, squash, mustard greens, and turnips.[69] The simple act of planting these garden seeds challenged the plantation system. In the past, many planters had wanted "every foot of ground" on sharecropper tracts "planted to cotton—'Cotton to the doorstep' "—and they had wanted "the tenants to buy food from the plantation stores." By planting gardens, croppers not only improved their diets, but they also reduced their dependence on the plantation stores that sold overpriced provisions on credit.[70]

In addition to furnishing tenant families with garden seeds, the Red Cross also loosened cotton's grip on the Southern bottomlands. By distributing soybean, alfalfa, and lespedeza seeds, the Red Cross reawakened interest in agricultural diversification. Although the soybeans that the Red Cross furnished in 1927 failed in the drought, the Red Cross committees began distributing soybean, alfalfa, lespedeza, corn, and cotton seeds in 1928 to flood-damaged communities. These counties and parishes received enough seeds to plant 342,992 acres of cotton, 243,908 acres of corn, 20,590 acres of alfalfa and lespedeza, and 66,225 acres of soybeans.[71] Although the acreage of soybeans, alfalfa, and lespedeza was rather modest in 1928, these crops were the harbingers of agricultural change in the Southern bottomlands (see chapter 5).

Soybeans had long been grown in the Southern states, but this valuable legume had been overshadowed by cowpeas. But during the early twentieth century, soybeans began supplanting cowpeas in the Southern cotton-belt. Both soybeans and cowpeas were nitrogen-fixing legumes that improved soils, but soybeans proved more versatile than cowpeas. Because many soybean varieties grew upright, it was easier to mow erect soybean plants for hay than it was to mow viny cowpeas. After mowing, soybean hay proved easier to cure than cowpea hay. Soybeans even had greater feeding value because soybean hay contained more seeds than cowpeas. Finally, soybeans were more prolific than cowpeas. A single acre of alluvial land, which was sown broadcast with soybeans, yielded two to three tons of hay.[72] Soybeans were not only planted as a hay crop to feed cattle, but were also grown in fields to fatten hogs. Because soybeans produced soft, fatty pork, bean-fed hogs had to be finished with corn in order to harden the pork. A common practice was to plant soybeans and corn together in the same fields and then turn in the hogs to "harvest" the crops. The hogs did all the work, consuming the beans first and then the corn. The beans fattened the hogs, and the corn firmed their flesh. Thus, "hogging-off" soybean–corn fields was an efficient, labor saving way of finishing hogs for market. After hogs gorged on beans and corn, farmers plowed under the litter and manure in order to fertilize the fields. Cotton was often planted the next season on these soybean–corn fields.[73]

Although the soybean proved a versatile field and hay crop, it was surpassed in productivity by alfalfa—the "king of hay crops." If alfalfa was seeded in the fall, it yielded an abundant hay crop the next year. It was possible to mow alfalfa fields three to six times during the growing season. Because each cutting yielded a ton of hay per acre, an alluvial alfalfa meadow produced three to six tons of hay per acre each season. Alfalfa not only furnished nutritious hay for cattle, but it also sustained hogs. A single acre of alfalfa supported more than a dozen hogs for most of the growing season. If stock raisers prevented overgrazing by hogs, it was possible to maintain a stand of alfalfa for several years. A properly maintained alfalfa meadow lasted five to seven years on bottomland soils. After this time, it was possible to plow up the alfalfa and plant cotton that thrived on the enriched soil.[74]

Lespedeza, or "Japan clover," was another versatile and productive legume. On upland soils, lespedeza grew low and spreading, furnishing little more than grazing for cattle. But on bottomland soils, this legume grew upright, allowing the plant to be mown for hay. Each season, bottomland lespedeza meadows yielded one to three tons of hay per acre. If sown as a pasture crop, lespedeza furnished grazing from early summer until first frost. Because it was a self-seeding crop, lespedeza, when mixed with grasses, provided permanent pastures for cattle. The D & PL estate, for example, transformed more than three thousand acres of sand-covered bottomlands into permanent lespedeza pastures after the 1927 flood.[75] Because lespedeza was a soil-enriching legume, it could also be used as a field crop as well as a pasture and meadow crop. It was possible to plant corn, then sow oats, and follow this with lespedeza. Cotton could be added "to this rotation after the lespedeza and before the land is again in corn."[76]

The lespedeza, alfalfa, and soybean seeds, which the Red Cross committees distributed, held great promise for the Southern bottomlands. First, these legumes added nitrogen to the soil, thus enriching lands where cotton was later grown. Second, these legumes were labor-saving crops, and this proved crucial after the 1927 flood when many Black laborers fled the bottomlands. Third, and most important, these legumes yielded an abundance of forage for cattle and hogs, thus providing the subsistence base for a renewed livestock industry.

The 1927 flood had depleted the livestock populations of the Southern bottomlands. During the flood, Red Cross workers rescued and fed 271,000 head of stock during the crisis. Yet, the flood drowned an estimated 9,300 work animals, 26,000 cattle, 128,000 hogs, and 1 million poultry. The Red Cross simply lacked the resources needed to replace all these lost animals. Red Cross committees provided replacement livestock and poultry to only 6,474 families who lost animals in the flood. They made a special effort to provide work animals, poultry, and milk cows, but the committees replaced only a handful of the lost cattle and hogs.[77]

Despite heavy livestock losses during the flood, rising meat prices reawakened interest in livestock raising. By 1929, the farm price for live cattle and hogs was nine cents a pound. Encouraged by improving prices, some planters borrowed funds to replenish their livestock herds. As an example, the number of cattle in flood-stricken Concordia Parish, Louisiana, increased from 4,154 in 1925 to 7,857 by 1930, and the number of hogs nearly doubled, growing from 10,919 to 18,145 during the same period.[78] In Concordia Parish, thousands of "high grade cattle" thrived on "luxurious pastures" of "white Dutch clover, lespedeza and Bermuda grass," and the parish annually shipped thousands of hogs to regional markets.[79]

Such successful examples of livestock raising, nevertheless, did not dissuade many bottomland planters from returning to cotton. Despite suffering enormous flood damage, the plantations of Washington County, Mississippi, increased the cotton land from 130,473 acres in 1925 to 147,003 acres by 1930. Although the farm price of cotton was only 17 cents per pound in 1929, many planters were still convinced that cotton was the most reliable money crop.[80]

Despite low cotton prices, it was possible to make money with careful management. The D & PL Co., for example, reversed its financial fortunes after 1927, thanks to the efficient administration of Oscar Johnston. Replacing Lant Salsbury as D & PL manager in 1927, Johnston's first task was to rescue the estate's Black tenants from the great flood. His second task was to save the fortunes of the flood-wrecked D & PL estate. Foreseeing a flood-induced cotton shortage, Johnston bought cotton futures on the New York Cotton Exchange. Buying futures in the spring, he liquidated his holdings in the fall of 1927. By playing the cotton futures market, Johnston earned two hundred thousand dollars for the rehabilitation of the wrecked D & PL estate. Despite the destructive flood, Johnston was determined to plant a cotton crop in 1928. After a string of unprofitable years under Salsbury's management, he was also determined to show a profit for the D & PL estate in 1928. Assembling the remaining tenants on the D & PL estate, Johnston demoted the share renters, who had paid one fourth of their crops as rent, to sharecroppers, who worked for half of the crop. By demoting renters to croppers, he not only increased the company's share of cotton, but Johnston also gained control over the estate's work animals. He assembled all the mules in central barns, where they received expert care from stock minders. Johnston believed "with expert supervision of the croppers (and the mules and tools)" that he could produce more cotton and earn higher profits for the company. He was indeed correct. When the accounts were settled in 1928, the D & PL estate turned an operating profit of $269,000.[81]

As Johnston was rebuilding the fortunes of the D & PL estate, the federal government began reconstructing the levee system in the lower Mississippi River Valley after the 1927 flood. The first priority was to repair about forty breaks in the valley levee line before the 1928 flood season. During the waning months of 1927, the U.S. Army Corps of Engineers and the local levee districts feverishly repaired the levee line.[82] When the Speakmans floated down the Mississippi River in a houseboat after the 1927 flood, they observed this repair work at Friar's Point, Mississippi, where "a government dredge boat" was rebuilding the damaged riverside levee.[83]

Although the repaired levees withstood the spring rise of 1928, the repair effort pushed local levee districts even deeper into debt. The total indebtedness of the levee districts in the lower Mississippi River Valley was estimated at $820 million. Realizing that the federal government would have to shoulder the burden of levee work, Congress passed the Jones–Reid Act in May 1928. Following the passage of this act, the federal government assumed the responsibility for flood control work in the lower Mississippi River Valley. The local levee districts were to provide rights of way for construction work, but the federal government was to pay the entire cost of flood control projects. The new flood control act not only provided for stronger levees, but it also called for floodways to relieve flooding in the Mississippi River channel. Allocating $24 million in 1928 for levee construction, the act appropriated $30 million annually after 1928 for flood control works in the Mississippi River Valley. Thanks to the Jones–Reid Act, levees became higher and stronger than ever before. The levees built before 1928 contained about four hundred thousand cubic yards of earth per mile, but the levees built after 1928 contained one million. Even the dangerous flood of 1929 failed to break these rebuilt levees.[84]

The threat of flooding was abating, but the cotton growers of the Southern bottomlands were facing an uncertain economic future. In 1929, when cotton was selling for seventeen cents a pound, the resurgent D & PL Co. earned an impressive $206,000 in operating profits. But in October of 1929, the stock market crash touched off the worst economic depression in U.S. history. During 1930, the farm price of cotton fell to ten cents per pound. And in that year, the D&PL Co. ran a $250,000 deficit.[85]

As cotton prices plummeted, cotton planters searched for ways to cut production costs, so they could still raise cotton profitably. One way to cut production costs was to substitute tractor power for human labor and mule power. Thanks to the introduction of the Farmall row-crop tractor during the 1920s, it was now possible for planters to use tractor-drawn equipment to plow, plant, and even cultivate cotton land.[86]

As an example, Johnie [*sic*] Brown, a Black farmer, who operated a fourteen-hundred acre Mississippi Delta plantation in partnership with a White merchant, bought a tractor on credit, so he could save labor costs in cotton planting. With "his tractor and other implements obtained one by one," Brown "put headlights on his machine and worked at night." When he "could get only a two-row [cotton] planter," Brown "fastened two planters abreast and made the first four-row planter in the neighborhood." Working his Issaquena County estate with tractors, Brown was among the pioneering Delta planters who mechanized their cotton estates.[87]

By 1930, a "few planters" in the Mississippi Delta had "given up the tenant system" and were "raising their cotton almost entirely by tractor equipment." Yet, even these mechanized planters needed some seasonal wage workers to "chop" (thin and weed) cotton plants in the spring and to pick the ripened cotton bolls in the fall. Fortunately, these planters lived "fairly close to towns" from which they obtained "labor for harvesting and chopping."[88]

Those fortunate planters who used seasonal wage workers and tractor-drawn equipment were able to produce cotton more cheaply than those who relied on tenants and mules. In 1930, the Delta agricultural experiment station at Stoneville conducted a convincing study, which compared production costs on mechanized plantations and on tenant plantations. M. G. Vaiden, a plantation auditor, compared the records from five tractor-operated Delta plantations with those from five tenant-operated plantations. Vaiden found that it cost only $35.61 on the average to produce an acre of cotton on a tractor plantation, whereas it cost $60.56 to produce an acre of cotton on a tenant-plantation. The difference in production costs was attributed to the high cost of housing and supporting the tenant families and the work mules on tenant-operated plantations. The conclusions of this study were obvious. If planters wanted to raise cotton profitably, they should dispense with tenants and mules and substitute tractor-drawn equipment.[89]

The adoption of tractors, however, involved a considerable capital investment. By 1930, the typical tractor cost $1,400 retail, and the cost of tractor-drawn implements added to this basic price.[90] Cotton prices were falling, and the economy was stagnating, thus planters found it difficult to borrow the capital needed to purchase tractors and tools. Following the stock market crash, banks began failing across the South, and the surviving banks were reluctant to loan money to struggling cotton planters.[91]

Even if planters managed to mechanize their cotton estate, there remained the problem of picking cotton. The picking of seed-cotton was the most laborious aspect of cotton production. From August to January, pickers dragged sacks up and down the rows, plucking the seed-cotton from the bolls. It required about five man days of labor to pick a single acre of cotton.[92] The only labor-saving solution to this problem was the development of a mechanical cotton picker. If mechanical cotton pickers were available, then "hand labor" on plantations could be "reduced to a minimum through the use of labor-saving machinery in all production operations."[93]

The quest for a mechanical cotton picker had begun in the nineteenth century. Between 1850 and 1930, inventors patented hundreds of designs for mechanical cotton pickers. Among these devices were threshing machines, pneumatic pickers, cotton strippers, and spindle pickers. The spindle pickers, which used rotating spindles to extract seed-cotton from bolls, were the most promising. During the 1920s, John and Mack Rust devised an effective spindle picker, which they patented in 1928. When tested at the Delta agricultural experiment station in 1933, the Rust machine managed to pick five bales of cotton a day. Despite this success, the Rust brothers failed to secure funding to manufacture their machine. It was not until 1942 that International Harvester developed a commercial cotton-picking machine for the Southern market.[94]

Because no mechanical cotton picker was available before 1942, cotton growers had to rely on human laborers to pick seed-cotton from the bolls. If planters lived near towns and cities, they sent trucks to collect day laborers to pick cotton. But if they lived in the remote countryside, planters had to rely on resident tenant families to pick their cotton crops.[95]

Although tractors did not eliminate the need for human labor in cotton picking, these power machines were well adapted for raising hay and forage crops. In Concordia Parish, Louisiana, for example, farmers used tractors to raise alfalfa, lespedeza, oats, corn, and soybeans. Using tractor-drawn equipment, it was possible to harvest as well as sow alfalfa, lespedeza, and oats. By planting corn in checks, it was even possible to cultivate the crop with tractor-drawn four-row cultivators. Yet, the most promising tractor-raised crop was the versatile soybean. By 1930, there was a growing market for soybean seeds. Since World War I, Southern farmers had been raising soybean seeds for the industrial oil market. When crushed, seeds yielded soybean oil, which was used for soap and paint, as well as soybean meal, which was excellent livestock feed. But in the 1920s, scientists developed a refining process to produce palatable soybean oil. Hydrogenated soybean oil could now be used for margarines and shortening. And in 1930, the Staley Company sold the first tank of refined soybean oil to a margarine manufacturer. Given the increased demand for soybean seeds, Concordia farmers used tractor-drawn implements to plant, cultivate, and harvest this labor-saving crop.[96]

Raising crops with tractors allowed Concordia's planters to overcome labor shortages during the agricultural recession. In the years between 1920 and 1930, the Black population of Concordia fell from 9,825 to 9,185. During the 1920s, hundreds of Concordia Blacks had joined the exodus from the South.[97]

The Black exodus from the South, which had begun during World War I, accelerated during the 1920s. The agricultural recession after the war dampened the hopes of many black Southerners. In addition, the boll weevil continued its advance across the South, reaching virtually all of the cotton South. The great flood of 1927 also added to the despair of Black residents of the Southern bottomlands. Finally, racial violence and brutal lynchings contributed to the exodus. In contrast to the troubled South, where an agricultural recession gripped the countryside, the industrial economy of the North prospered during the 1920s, and the Northern industrial cities beckoned to dispirited Southern blacks. Given the relative prosperity of the industrial North and the comparative poverty of the agricultural South, it was not surprising that an estimated 1.4 million Blacks left the Southern states during the 1920s.[98] In addition to this emigration from the South, there was also an internal migration from the countryside to the Southern cities. As a result of these migrations, the Black rural-farm population of the South declined dramatically during the 1920s. Yet, this decline was not shared equally among the Southern states. The Southeastern states suffered the greatest losses, whereas the Southwestern states suffered lighter losses or even experienced gains. As example, the Black rural-farm population of Arkansas declined by 3 percent during the 1920s, but the Black rural-farm population in Mississippi grew by 10 percent during the same decade.[99]

Despite recessions, weevils and floods, the Mississippi Delta attracted rural-farm Blacks from other Southern states. Black families from the upland South, whose cotton crops had been devastated by boll weevils, sought farms and jobs in the Mississippi Delta.[100] Yet, not all the Delta newcomers were Black. Many hill-country Whites, who were derisively called " 'rednecks' or 'peckerwoods,' " joined the stream of immigrants to the Mississippi Delta, where they competed with Black residents for farms and jobs.[101]

Immigration to the Southern bottomlands continued into 1930—the year that the agricultural recession ended and the "Great Depression" began. During this crucial year, the federal government completed the most thorough census of the U.S. people to that date. The massive volumes of the Fifteenth Census, which were devoted to population and agriculture, presented a detailed portrait of the countries and parishes of the Southern bottomlands. The census data from these volumes revealed the startling changes that had occurred in the Southern bottomlands since 1880.

In the years between 1880 and 1930, the Southern bottomlands evolved from an "extensive agricultural frontier" with fewer than eighteen people per square mile into an intensive agricultural region with more than forty-five people per square mile.[102] In 1880, the countries and parishes of the Southern bottomlands counted only sixteen people per square mile. By 1930, the population density of the bottomland communities was an impressive fifty-one people per square mile (see Tables 1–2).

In just five decades, the population of the Southern bottomlands had tripled, the acreage of tilled land had nearly quadrupled, and cotton production had more than tripled. Between 1880 and 1930, the population of the bottomland communities soared from 390,014 to 1,221,500 people. Blacks were 61 percent of the bottomland population in 1880, but they were only 50 percent of the population by 1930. During the years between 1880 and 1930, the acreage of tilled land rose from 1,708,546 to 6,383,699 acres. Bottomland residents planted only 45 percent of the tilled land in cotton in 1880, but they cropped 57 percent of the tilled land in cotton by 1930. Strengthening its hold on the Southern bottomlands, cotton remained the most valuable commodity. Between 1880 and 1930, the production of cotton lint increased from 579,927 to 1,971,989 bales (see Tables 3–8).

After five decades of settlement, the Mississippi Delta had surpassed the Louisiana Delta as the most densely settled area in the Southern bottomlands (see chapter 2). Counting 418,003 inhabitants, the Mississippi Delta had a population density of 69 people per square mile. Unlike the Arkansas and Missouri Deltas, where Whites outnumbered Blacks, the Mississippi Delta had a Black majority. Most of the Mississippi Delta's 306,813 Black residents lived and worked on the area's plantations and farms. By 1930, the Delta's plantations and farms tilled 2,026,483 acres or 52 percent of the land in the Delta. Planting one fourth of the tilled land in corn and other crops, Delta agriculturists devoted three fourths of the cultivated land to cotton. In 1930, Mississippi Deltans planted 1,534,389 acres in cotton, which yielded 873,659 bales (see Tables 1–8).

The Mississippi Delta produced about four times as much cotton as the Louisiana Delta. Planting 406,859 acres in cotton, Louisiana Deltans harvested only 222,274 bales from their weevil-ridden fields. The Louisiana Delta not only produced far less cotton than the Mississippi Delta, but the area also counted far fewer residents, and it had a much lower population density. Counting 153,006 inhabitants in 1930, the Louisiana Delta contained only 32 people per square mile. Despite the Black exodus of the 1920s, most of the Louisiana Delta's inhabitants were Black, and the bulk of the Delta's 82,328 Black residents lived on plantations and farms. By 1930, the Delta's plantations and farms tilled 703,613 acres or only 22 percent of the total land. The remaining land in the Louisiana Delta was unimproved farmland, cut-over timberland, and raw swampland (see Tables 1–8).

Much of the Arkansas Delta was also untilled after decades of settlement. By 1930, the residents of the Delta counties tilled 2,406,334 acres or less than 40 percent of the land in the Arkansas Delta. Counting 479,019 inhabitants, the Arkansas Delta had a population density of 49 people per square mile. In contrast to the Louisiana and Mississippi Deltas, where Blacks outnumbered Whites, the Arkansas Delta counties held 282,012 Whites or 59 percent of the population. Although some of the White residents were townspeople, most resided on Delta plantations and farms. By 1930, the plantations and farms in the Arkansas Delta cropped 1,335,041 acres of cotton and harvested 662,019 bales (see Tables 1–8).

Although agriculturists in the Arkansas, Louisiana, and Mississippi deltas planted most of the tilled land in cotton, farmers in the Missouri Delta cropped more acreage in corn than in cotton. Cropping 382,332 acres in corn, they planted 330,029 acres in cotton. This modest cotton acreage, nevertheless, yielded 214,037 bales or two thirds of a bale per acre. Despite the short growing season, the relatively weevil-free fields of the Missouri Delta actually produced more cotton per acre than the weevil-infested fields of the warmer Arkansas, Louisiana, and Mississippi deltas. By 1930, Missouri Deltans planted one fourth of the tilled acreage in cotton. Much of this cotton land was cultivated and harvested by Black tenants and workers, who had moved up from the lower South. Attracting Southern Blacks as well as Southern and Midwestern Whites, the Missouri Delta counted 171,472 residents, and the population density surpassed 52 people per square mile. By 1930, agriculturists tilled 1,247,269 or 60 percent of the land in the Delta (see Tables 1–8).

Throughout the Southern bottomlands, agriculturalists tilled 42 percent of the land by 1930 and they owned 55 percent of the bottomland acreage. In 1880, agriculturalists tilled only 11 percent of the land, and they owned only 31 percent of the acreage in the Southern bottomlands (see Tables 5-6). Between 1880 and 1930, much of the acreage in the Southern bottomlands passed from the hands of governments, corporations, and speculators into the hands of planters and farmers (see chapters 2–4).

Acquiring acreage from public and private owners, agriculturalists who settled in the Southern bottomlands confronted a host of environmental problems, including periodic floods, swampy soils, dense vegetation, physical isolation, insect pests, and endemic diseases. Yet, between 1880 and 1930, the residents of the Southern bottomlands largely overcame these environmental hindrances. Thanks to federal aid, local levee districts replaced feeble private levees with massive earthen embankments that held back floodwaters in most years. Although the levee line failed during the 1927 flood, the U.S. Army Corps of Engineers rebuilt the levees, and these embankments withstood the 1929 flood. In turn, local drainage districts turned thousands of acres of cut-over and timbered swampland into reclaimed farmland. The drainage of mosquito-ridden swamplands not only yielded farmland, but it also reduced the threat of such mosquito-borne diseases as malaria. After draining, workers and tenants transformed the reclaimed swamplands into cleared fields by removing the vegetation. After partial clearing, it was possible to use mule-drawn plows to cultivate fields until the stumps decayed, but the use of tractors required more thorough clearing to remove these obstructions. As the swamplands were drained, cleared and farmed, a growing network of highway and railways overcame the physical isolation of the bottomlands, allowing agriculturalists to send cotton products to regional markets. Receiving aid from federal agencies, bottomland agriculturalists learned to combat the boll weevils that infested cotton fields, and they overcame the livestock diseases that had prevented stock improvement. Finally, bottomland residents learned to cope with the threat of malaria by taking quinine and by curbing malaria-bearing mosquitoes. Overcoming the environmental hindrances that had stymied the early settlers (see chapter 1), the agriculturalists of the Southern bottomlands transformed much of the forbidding swampland into productive farmland by 1930. Yet, this achievement was not accomplished without cost. During the years between 1880 and 1930, hundreds of thousands of acres of swampland forest, which would later be regarded as valuable "wetland habitats" by ecologists, were drained, cleared, and farmed.[103] This destruction of hardwood forests occurred at an alarming rate during the early twentieth century. As an example, the parishes of East Carroll, Madison, and Tensas in the Louisiana Delta still held about 459,000 acres of hardwood timberland in 1917. Thirteen years later, these parishes contained only 215,000 acres.[104]

In addition to the loss of wetland habitats, there was also a human cost. The Southern bottomlands became a region of "rich land" but "poor people."[105] By 1930, the vast majority of bottomland farmers did not own the land that they farmed. Landless tenants (sharecroppers, sharerenters and cashrenters) operated 88 percent of the farms in the Southern bottomlands. Fifty years earlier, landless tenants operated only 53 percent of the farms in the bottomlands. The 1930 census revealed that tenants worked 95 percent of the farms in the Mississippi Delta, 88 percent of the farms in the Louisiana Delta, 83 percent of the farms in the Arkansas Delta, and 80 percent of the farms in the Missouri Delta (see Tables 9–10). In the years between 1880 and 1930, the rural-farm population of the Southern bottomlands became a mass of poor Black and White tenants. These tenants worked for the well-capitalized planters and investors who owned the bulk of the farmland in the Southern bottomlands. In the largely Black Mississippi and Louisiana deltas, most of the tenant farmers were Blacks. The D&PL estate in the Mississippi Delta, for example, worked its Scott plantation with almost one thousand Black cropper families.[106] But in the largely White Arkansas and Missouri deltas, most of the tenant farmers were Whites. As an example, Abner Driver of Osceola, Arkansas, rented his plantation to about 230 White tenant families.[107] Working on their "tenant tracts," croppers and renters raised most of the cotton in the Southern bottomlands, but they failed to share in the agricultural rewards. Tenant families lived in rickety shacks, which barely kept out the wind, rain and mosquitoes, they survived on coarse foods, their children rarely received a decent education, and tenant families suffered from a host of diseases.[108] Living in dismal poverty, Black and White tenant families feared the worst as the first shocks of the Great Depression struck the Southern bottomlands in 1930, pushing the price of agricultural commodities to record lows.

5

Settling the Southern Bottomlands After 1930

> In a day of increasing concern over problems arising from the disappearance of the old land frontier, the Nation, long accustomed to looking westward in search of new land opportunities, is now witnessing a major tide of new settlement in the Mississippi Delta [lower Mississippi Valley], one of its oldest and most fertile farming regions.
>
> More than 20,000 farm families, most of them settling upon small, family-size tracts, have gone into the undeveloped areas of the Mississippi Delta [lower Mississippi Valley] in the last decade and are clearing land and attempting to develop new homes and farms.[1]

As the agricultural recession of the 1920s ended and the Great Depression (1930–1941) began, crop prices plunged to the lowest levels since the depression of the 1890s (see chapter 2). In 1920, the farm price of cotton lint—the best indicator of the economic health of the Southern bottomland—stood at seventeen cents per pound. By 1931, the price of cotton lint tumbled to six cents per pound. The bottomland tenants, farmers, and planters, who depended on cotton for their livelihood, found their incomes slashed to a fraction of 1929 levels.[2] Not only had the price of cotton fallen to record lows, but the onset of the Great Depression coincided with a terrible drought that parched the Southern bottomlands.

The great drought began in the Midwest and upper South during the spring of 1930. Between March and May, the spring rainfall was the lowest on record in the state of Missouri. During the summer, the drought shifted to the lower South. The rainfall for the months from June to August was the lowest ever recorded in Arkansas, Mississippi, Louisiana, and neighboring states. After weeks of such hot, dry weather, the spring-planted crops withered and died in a score of states.[3]

Because the drought destroyed crops over such a large area, it became a national agricultural crisis by the late summer. Responding to the crisis, President Herbert Hoover, who had directed the flood relief effort in 1927, called a conference of governors from the drought-stricken states in August. The conference agreed to establish a federal drought relief committee, state drought committees, and even county committees to deal with the crisis. The newly created federal drought committee called for federal and state loans to provide crop seed for the drought-stricken farmers, but it overlooked the need for food and forage in the drought-damaged areas. The drought not only killed much of the cotton crop, but it also withered the cornfields, gardens, meadows, and pastures. In many areas, little food and forage was left to sustain farm families and their livestock over the winter months. President Hoover was opposed to federal "doles" to drought victims, so he called on the American Red Cross, the national charity which had aided victims of the 1927 flood, to provide relief to the 1930 drought sufferers.[4]

Although the Red Cross pledged $5 million from its disaster reserve fund for drought relief, Red Cross Chairman John Payne was reluctant to provide direct relief to drought victims. He preferred such "self-help" projects as planting fall gardens and sowing winter pastures. Under Payne's direction, local Red Cross chapters distributed packages of garden and pasture seeds to families in droughty areas. In Arkansas, for example, the Red Cross handed out thousands of packages of vegetable and pasture seeds to needy farm families. When hungry farmers dutifully planted these seeds, most of the crops failed to survive the drought and heat. Often the only vegetables which survived were greens and turnips. Arkansans dubbed the turnips, "Hoover apples," and they joked about eating "turnip sandwiches"—a slice of turnip between two other slices of turnips. Because the fall garden and winter pasture project failed to feed the hungry people and starving livestock, the local Red Cross chapters would have to furnish relief aid to drought-stricken counties and parishes.[5]

In some bottomland communities, the planters, who often controlled the local Red Cross chapters, feared that seed packages and Red Cross rations would "demoralize" their tenants. They wanted the hungry share croppers to pick the surviving cotton for reduced wages. With no garden crops and no relief rations, desperate "croppers had no choice but to work for less wages, even if they were hungry and weak."[6]

But after the cotton was picked, even the most hard-hearted planters admitted the need for drought relief. Thanks to searing drought and falling prices, the 1930 cotton crop was worth only a fraction of previous crops.

In Arkansas, for example, the 1930 cotton crop earned only $90 million. The year before, the cotton crop had earned $210 million.[7]

The poor cotton crop of 1930 proved disastrous for the fragile plantation credit system. In normal years, planters borrowed funds from bankers in order to finance their forthcoming crops. Planters, in turn, "advanced" money and supplies to their tenants, so they could plant, cultivate, and harvest the cotton crop. After the cotton was harvested, the tenants repaid the planters for these advances, and the planters in turn repaid their loans to the banks. But in 1930, the poor cotton harvest broke this cycle of credit. Tenants could not repay planters, and planters could not repay bankers. Many banks, which were barely solvent, began to close their doors in late 1930. In the month of November, more than one hundred banks failed in the state of Arkansas. Unable to borrow funds to put in the 1931 cotton crop, many bottomland planters lacked the resources to care for their destitute tenants after cotton-picking season ended.[8]

As the drought famine worsened, congressmen from the drought-ravaged states called for massive federal aid to help the drought victims. Yet, when Congress tried to pass a bill providing $60 million in federal loans to drought-stricken farmers, the Hoover administration balked at the figure. Hoover's budget director proposed only $25 million in federal loans. Finally, in December 1930, Congress compromised on a $45 million farm loan bill. Although this bill would loan federal money for crop seed and for stock feed, there was no provision for food loans. The federal government left the problem of feeding drought victims to the Red Cross.[9]

Despite Payne's reluctance to provide direct relief to drought victims, it was apparent by November 1930, that "thousands would be dependent upon Red Cross aid in the coming winter." In December, local Red Cross chapters began distributing food, clothing, and fodder on "a basis of need." The Red Cross, however, did not simply hand out tons of supplies to queues of drought victims. In dispensing relief, the organization followed a strict bureaucratic procedure. Relief committees were set up by local Red Cross chapters oversee the distribution of relief aid. Investigating committees went into the countryside to canvass needy families. Interviewing drought victims and filling out questionnaires, investigators returned those forms to the relief committees, which issued disbursement orders to drought sufferers. Taking these disbursement orders to local merchants, drought victims received food, clothing, and even fodder. The merchants, in turn, presented their bills and orders to chapter treasurers for reimbursement. If the process went smoothly, drought victims received timely aid form their local Red Cross chapters.[10] But all too frequently, bureaucratic snafus hindered Red Cross efforts to aid victims of the 1930 drought.

In Lonoke County, Arkansas, for example, the Red Cross chapters were trying to feed hundreds of destitute tenant farmers on local plantations. The committees, however, used up their Red Cross forms by December 31, 1930. Requesting additional forms from the Red Cross offices, the chapters were unable to issue more rations to needy families until they received the necessary forms. Angered by the interruption of Red Cross aid, a group of hungry tenants, who were led by a White tenant named H. C. Coney, headed into England, Arkansas, on January 3, 1931. Gathering in front of a grocery store, they demanded food for their hungry families. Soon hundreds of tenants and townspeople were clamoring for Red Cross rations. Fearing violence, G. E. Morris, the chairman of the Lonoke County drought committee, promised the people aid. England's merchants furnished about fifteen hundred dollars worth of rations to some five hundred families. Receiving sacks of provisions, the protesters returned peacefully to their homes.[11] Yet, Coney admitted that the crowd could have turned violent: "It wouldn't a wanted but just a little bit o' sass ter've had a showdown. But they doled out the feed and we all rolled back here without nobody gettin' hurt."[12]

This peaceful food protest in England, Arkansas, which was called a "food riot" in the newspapers, drew national attention to the plight of Arkansas' drought sufferers. News of the England food protest soon reached Washington, D.C., where it prompted action in the U.S. Congress and in the Red Cross headquarters. On January 5, 1931, Senator Thad Caraway of Arkansas introduced an amendment to the recently passed $45 million farm loan, adding $15 million for food loans. Although Hoover's Congressional supporters voted down Caraway's amendment, Senator Joseph Robinson of Arkansas then sponsored a bill to allocate $25 million for the Red Cross to be used for drought and unemployment relief. Although Hoover's supporters quashed this bill as well, the American Red Cross realized the need to raise more funds for drought relief.[13]

On January 10, Payne called on U.S. citizens to donate $10 million to the Red Cross for drought relief. By the end of December 1930, the Red Cross had spent $446,000 of the disaster reserve fund, and the local chapters had spent an additional $404,000. Because local chapters had exhausted their funds, and because drought relief had reduced the national disaster reserve fund, Payne launched a national fund-raising effort. As Americans responded to Payne's appeal, the Red Cross coffers filled with more than $10 million in contributions for drought relief. In addition to money, Americans also contributed surplus food, used clothing, and livestock fodder to the drought relief effort. Thanks to this charitable outpouring, the Red Cross was aiding more than 460,000 families in the drought-damaged states by the end of February 1931.[14]

Arkansas, which suffered the worst drought damage of any state, received most of the Red Cross relief aid. Nearly 160,000 farm families in Arkansas accepted Red Cross food, clothes, fodder, or medical care.[15] Yet, the recipients of Red Cross relief did not live in leisure and luxury. In Poinsett

County, Arkansas, for example, the recipients of Red Cross rations had to work at "clearing and grubbing" plantation land in return for slips entitling them to one dollar's worth of Red Cross supplies for a day's work. The work was worth two or three times as much. Although public outcries ended Poinsett's "works program," the disbursement of Red Cross relief remained in the hands of local planters, who dominated the relief committees in bottomland communities.[16] Fearing that tenants would be spoiled by "rich food," relief communities disbursed the plainest sort of rations. As an example, in St. Francis County, Arkansas, a family of five was expected to live on five dollars worth of groceries each month. These rations included flour, cornmeal, cracked rice, split beans, lard, molasses, coffee, and baking powder. Recipients of Red Cross rations were so desperate for meat that they hunted rabbits, or "Hoover hogs," in the swamps, and they stole chickens from their more fortunate neighbors.[17]

Although Red Cross aid was often deficient in quality and quantity, the rations, clothes, medicines, and fodder, which the organization disbursed during the winter of 1931, did save numerous people and animals. But in the spring, the Red Cross began shifting from relief aid to self-help projects. During March and April, the Red Cross distributed packages of garden and pasture seeds to drought victims. In contrast to the ill-fated seed distribution effort in the fall of 1930, which reached only 86,000 families in the drought-ravaged states, the seed distribution program in the spring of 1931 reached 605,000 families. In Arkansas, more than 123,000 families received packages of garden and pasture seeds in the spring of 1931. When planted, these seeds were watered by the spring showers that finally broke the devastating 1930–1931 drought.[18]

Gardens and pastures helped feed families and livestock during the 1931 season, but agriculturalists also needed credit to put in their crops of cotton and corn. In ordinary times, planters borrowed funds from bankers to furnish their tenants with cash and supplies. But in 1931, the surviving banks had "too many frozen loans to advance new credit." Although some banks reopened, bankers were reluctant to make loans to planters who were "bankrupt or nearly so."[19] In order to relieve local banks from the task of farm financing, the Arkansas state legislature appropriated $1.5 million toward the creation of county agricultural credit corporations. By June 1931, these credit corporations had loaned $750,000 to Arkansas farmers.[20] In addition to borrowing from agricultural credit corporations, Arkansas turned to the federal farm loan program, which Hoover had signed into law in January 1931. These farm loans offered funds for crop seed and for livestock feed. When applying for a federal loan, the borrower "submitted his plans for planting, stating the number of acres to go into feed crops and the number in cash crops," and "how much feed was needed for the workstock." As security for the loan, the borrower "gave a first mortgage on the crops to be planted."[21] Borrowers not only had to pledge forthcoming crops as collateral, but they also had to endure "endless red tape."[22] To administer the farm loans, the U.S. Department of Agriculture set up regional offices at Memphis, St. Louis, and elsewhere, and the department established local committees to rule on federal loan applications.[23] The local loan committees, which were composed of bankers, merchants, and planters, approved only the loans of applicants who were considered "respectable." Small farmers, who were heavily indebted to bankers and merchants, found it difficult to secure government loans. In addition, tenants could obtain farm loans only if their landlord "waived the lien on their crop, a rare favor."[24] Coney, the tenant farmer, who had led the England "food riot" in January 1931, applied for a $250 federal farm loan, but his landlord, a Memphis cotton broker, was reluctant to waive his lien on Coney's forthcoming crop. In May, Coney was still waiting for his farm loan, but he vowed "to make a crop," even if he had "to graze one mule and plow the other."[25] Because tenants and small farmers rarely qualified for farm loans, most of the federal seed and feed loans went to "respectable" planters. Yet, the federal government imposed a two thousand dollar limit on seed and feed loans, and this limitation made it difficult for planters to borrow enough money to properly furnish their tenants with supplies.[26] Despite all these shortcomings, the federal farm loans provided some needed credit for agricultural rehabilitation after the great drought. By June 1931, more than fifty thousand Arkansans had received about $8 million in seed and feed loans.[27] Thanks to federal and state loans, Arkansans scratched in their 1931 crops, and the spring showers brought "new hope to an adversity-ridden people."[28]

As Arkansans were recovering from the great drought, other areas counted their drought losses. Southeastern Missouri, for example, had been seared by months of rainless weather. In New Madrid County, not a single drop of rain had fallen for ninety days, and the countryside had been "reduced to a state of absolute poverty."[29]

In southeastern Missouri, the drought and the depression caused dozens of banks to fail, and a wave of foreclosures followed the bank failures. Among the farmers who experienced foreclosure was Thad Snow of Mississippi County, Missouri. Snow's bank sold his obligations along with those of others to a leading St. Louis financier in a vain effort to save itself. When the bank failed, the financier foreclosed on Snow, picking his "bones clean" and leaving him "without an acre of land or a mule or a plow." Refusing to accept defeat, Snow turned to the last seven solvent businessmen in Mississippi County. He borrowed the funds needed to reclaim his estate from the St. Louis financier. After Snow declared bankruptcy, his creditors deeded him the land, and Snow gradually repaid his benefactors for the loan. Landholding farmers were not the only ones to lose property. Jim Johnson, a successful tenant farmer in Mississippi County, Missouri, rented a section of land, which he worked with "a dozen good mules and the equipment to go with them." Johnson had mortgaged his mules and tools in order to finance his 1930 crop, but he had failed "to pay out." He still owed $600 on his note, and "his note fell into the hands of a St. Louis note-shaver," who planned to foreclose on Johnson. Fortunately, Johnson found a local merchant, Simon Loebe, who had just sold his billboard business, and Johnson convinced Loebe to loan him the money to pay off his debts. Saving his mules and tools from foreclosure, a thankful Johnson put in his 1931 crops on rented land.[30] Johnson and Snow managed to save their properties from foreclosure, but many renters and farmers had their tools, mules, and lands foreclosed and auctioned at the courthouse doors.

Although numerous renters and farmers lost their property during the drought, the poor share croppers suffered even greater privations. During the drought, desperate croppers were reduced to stealing corn from farmers' cribs. Taking a few ears of corn at a time, they parched corn kernels in skillets with a little lard and salt. Only Red Cross relief aid saved these destitute croppers from actual starvation. During the drought crisis, the Red Cross furnished relief aid to more than twenty-two thousand farm families in Missouri.[31]

Red Cross aid also proved crucial in Louisiana and Mississippi. During the drought crisis, the Red Cross provided relief aid to over 107,000 farm families in these states. Much of this relief went to the bottomland parishes and counties, where the drought had seared the 1930 crops.[32]

Unlike the 1927 flood, which had burst on the Louisiana and Mississippi deltas (see chapter 4), the 1930–1931 drought was "quiet and slow, without sensation or glamor."[33] In Louisiana, the drought ruined the "food and feed crops," and the "cotton yield did not promise more than a sixth of the normal crop in the northern parishes." Yet, the drought was only one factor in a very "lean year." Poor crop prices and limited bank credit exacerbated the drought crisis in northern Louisiana. In neighboring Mississippi, the residents also had "to contend with a combination of drought, a poor cotton crop with prices ruinously low, and a bank collapse." One third of Mississippi's banks failed during the drought crisis, leaving the credit system in shambles.[34]

Because the Mississippi state government, which was then under the dubious leadership of Governor Theodore Bilbo, did little to ease the crisis, the Red Cross had to care for thousands of destitute people. During the drought, many poor White people left the hill counties and moved into the bottomland counties, hoping to find jobs as cotton pickers. The bottomland plantations, nevertheless, had little work for these White transients, who lived in vacant shacks and who survived on Red Cross aid. Many Black share croppers in the Delta also subsisted on Red Cross relief, because so many planters could no longer afford to furnish supplies to their tenants.[35] In fact, some planters could not even feed themselves. Planters, "who after the World War were rated as men of wealth, were forced to apply to Red Cross local committees for relief, the same as their unemployed field hands."[36]

Although relief aid was disbursed rather efficiently by Mississippi's Red Cross chapters, the relief effort was not without controversy. In Bolivar County, Mississippi, for example, mules fared better than people. A work mule got $8 per month for fodder, whereas a hungry family of six people received less than $12 worth of groceries each month. Red Cross aid was not only stinted, but it was also brief. Red Cross relief work did not begin until January 1931, and local chapters began ending relief aid in March—the month that planters traditionally began furnishing their tenants.[37]

As Red Cross relief aid was ending, Mississippi's planters sought credit from federal seed and feed loan programs, agricultural credit corporations, and the surviving state banks in order to plant the 1931 crops. Because credit was so scarce, they were forced to plant one of the "cheapest" crops ever made. Bottomland planters raised more food and forage crops and less cotton than in previous years.[38] Even tenant farmers "raised a bountiful food crop" that year.[39]

Those food and forage crops proved invaluable during the 1931 season, when cotton prices sank to the lowest levels since 1898. In the local markets, cotton sold for only five or six cents a pound. A bale of cotton, which had brought $215 "during the flush years following the War," fetched about $27 during the fall of 1931. Not only did cotton lint prices fall to record lows, but the price of cotton seed plunged to less than $9 a ton.[40]

The drop in cotton seed prices was particularly disastrous for tenant farmers. On Southern plantations, it was customary for planters to give the cotton seed to their tenants. After deducting enough seed to pay for ginning their cotton, tenants sold the rest of the cotton seed. This "seed money" was often "the only real money" that many share croppers received from their cotton. After repaying their debts, many croppers were lucky to earn any money from their cotton crop—other than the seed money. When cotton seed sold for forty dollars to twenty dollars a ton, croppers had enough seed money left after ginning to buy their families some clothing, medicine, and food. Yet, when the price of cotton seed fell to less than ten dollars per ton, croppers barely paid for the ginning, and there was little money left over for their families. Unable to buy any new clothes since 1929, cropper families wore ragged, patched garments. In order to purchase new clothes and other goods, they had to have "a little cash on hand." To get cash, croppers would have to sell their "corn and peas for twenty-five cents a bushel," or they would have to steal.[41]

Among the cotton tenants who endured the hard times of the early 1930s were the Knights—a White family who share cropped on a plantation near Cleveland, Mississippi. Because the landlord insisted that they plant nothing but cotton on their tract, the Knights had to raise "their garden of tomatoes, okra, and cucumbers between the cotton rows." Enduring malaria and pellegra, the Knights dutifully planted, chopped, and picked cotton. Receiving "advances" of twenty dollars per month from March through August, the

Knights spent this princely sum on food and other essentials. After harvesting their cotton in the fall, they repaid the landlord for these advances. Keeping the cotton seed money after ginning, the Knights bought some food to tide them over until the following year.[42]

Falling cotton prices not only hurt tenants but also threatened planters. Bottomland planters had struggled financially since the 1927 flood. By 1930, many planters were deeply in debt. Unable to pay off their debts, they lived under the threat of foreclosure. Cash was so scarce that some planters failed to pay their land taxes, so they lost acreage to tax forfeiture.[43] Because bankers were reluctant to make loans to indebted, tax-delinquent planters, many struggling agriculturalists faced bankruptcy if cotton prices did not rebound. In 1931, when cotton prices hovered around five cents per pound, numerous estates were sold through tax delinquency, bankruptcy action, and mortgage foreclosures.[44]

Foreclosures accelerated in 1932, when the price of cotton stagnated at six cents a pound, and cotton seed fetched barely ten dollars a ton. In many ways, 1932 was the darkest year since agricultural prices collapsed in 1920. For twelve years, bottomland planters, farmers, and tenants endured poor prices, droughts, floods, and boll weevils. They kept hoping that prices would eventually improve, so they could relive the prosperity of the World War I era. But in 1932, their hopes were finally dashed.[45]

Despite the deepening depression, Southern planters, farmers, and tenants could expect little help from the federal government during President Herbert Hoover's term (1929–1933). Hoover's Republican administration was reluctant to offer federal aid to struggling agriculturalists. Hoover hoped to improve the farmers' lot through price stabilization and federal loans to credit-giving institutions. In 1929, Hoover signed the Agricultural Marketing Act, which established the Federal Farm Board. The board lent federal funds to marketing cooperatives, and established price stabilization corporations. Notable among these corporations was the Cotton Stabilization Corporation, which could buy, store, and sell cotton in order to stabilize cotton prices. Although the Cotton Corporation purchased 3.4 million bales by the end of 1931, the price of cotton kept falling. There was no way to control cotton prices without controlling production. In 1932, the Federal Farm Board called on cotton growers to plow up every third row of cotton. By voluntarily reducing cotton production, growers could drive up cotton prices. Yet, few farmers heeded the board's plea. They needed to raise as many bales as possible in order to offset the low cotton prices. Unless the federal government paid farmers to take cotton land out of production, growers could not afford to reduce cotton production. In addition to these failed attempts to stabilize prices, the Hoover administration tried to shore up credit-giving institutions, which lent funds to agriculturalists. In 1932, Hoover approved the creation of the Reconstruction Finance Corporation (RFC), which offered federal loans to struggling banks, trust companies, and other lending institutions. The RFC also made agricultural, commercial, and industrial loans. Arkansans, for example, received more than $4 million in agricultural loans during 1932. Yet, the RFC and the Cotton Stabilization Corporation failed to ease the agricultural depression. These federal efforts did not address the underlying problem in Southern agriculture: The South was producing more cotton than domestic and foreign markets could absorb. Because the markets were glutted, cotton lint sold for a few pennies a pound and cotton seed fetched a few dollars a ton. In order to raise the price of cotton products, it was necessary to reduce the supply by curbing cotton production.[46]

In September 1932, Franklin Roosevelt, the Democratic presidential candidate, espoused the principle of crop controls in order to raise commodity prices. Offering hope to America's beleaguered farmers, Roosevelt won the support of agrarian voters in the November election. Although Roosevelt won a decisive victory in November 1932, he could not assume office until March 1933. During the next four months, Hoover, the "lame-duck" president, did little to alleviate the worsening agricultural crisis.[47]

When Roosevelt was inaugurated in March, the U.S. cotton industry was in disarray. Not only was there a "carry-over" of 12.5 million bales from the 1932 season, which still glutted the world market, but U.S. growers planned to raise even more cotton in 1933. Encouraged by Roosevelt's promise to control crop production, speculators were buying surplus cotton, thus driving up the price from six cents a pound in March to eight cents by May. As cotton prices rose, U.S. growers planted more than 40 million acres in spring 1933, and this huge cotton crop threatened to swamp the world market. Hoping to reduce the surplus of cotton and other commodities, the Roosevelt administration drafted a crop-control bill as part of its flood of "New Deal" legislation. This bill allowed the federal government to pay farmers to withdraw land from production in order to control crop surpluses. When Roosevelt sent the bill to Congress in March, skeptics in Congress delayed its passage for weeks. When Congress finally passed the bill, Roosevelt signed the Agricultural Adjustment Act in May.[48]

The act permitted the creation of a new federal agency—the Agricultural Adjustment Administration (AAA)—to oversee the crop-control program. Within the AAA bureaucracy, there was a division of production that contained a "cotton section." Headed by Cully Cobb, the cotton section had to convince Southern cotton growers to reduce the 1933 crop in order to lessen the crop surplus and thus raise cotton prices.[49]

Because the 1933 cotton crop was largely planted when the AAA was created in May, the agency planned to plow up 10 million acres of cotton land in order to reduce the projected cotton surplus. On June 19, the AAA announced that it would contract with growers to plow up 25 percent to 50 percent of their cotton acreage. As compensation for plowing up cotton, growers could receive a cash payment of seven to twenty dollars an acre (depending on the expected yield), or they could choose to participate in the "cotton option" program. If the cotton option was chosen, the grower would receive a cash payment of six dollars to twelve dollars an acre, and would be able to buy an amount of government cotton (equal to the quantity destroyed) at the price of six cents per pound. If the market price went above six cents a pound, the grower could sell the government cotton, which had been stockpiled during Hoover's term, at the higher price and pocket the difference. To generate money for the plow-up payments, the AAA planned to impose a controversial processing tax on commodities. Offering cash payments and cotton options to cooperating growers, the AAA launched its cotton "plow-up" campaign during the summer.[50]

When the "word went out" that they "were to plow up every third row" of cotton, there was "a good bit of head-shaking" among the cotton growers. They were "destroying a crop that was almost ready for harvest," and they were "doing violence" to their "strongest tradition and habit." Even the mules were bewildered by the "plow-up" program. The "mules had been trained to guard and not to damage the cotton," and it was difficult to get them to plow up the rows.[51]

Although some growers remained skeptical about the "plow-up" program, the AAA signed up more than 1 million cotton growers, who plowed up 10.5 million acres of cotton. Plowing under millions of acres, U.S. growers reduced the cotton crop from an estimated 17 million bales to 13 million harvested bales. This reduction in the 1933 cotton crop contributed to a modest rise in cotton prices. Between 1932 and 1933, the price of cotton edged upward from six cents to ten cents per pound, and this price increase added an estimated $114 million to the cash value of the 1933 cotton crop. In addition, the growers who participated in the "plow-up" program received $112 million in cash payments, and the option on government cotton earned them another $48 million. Yet, this newfound wealth was not equitably distributed among the cotton growers. The landowning planters and farmers received the lions' share of the AAA payments, and the renters and croppers received only the scraps.[52]

Following "accepted Southern practice," only landlords signed the 1933 plow-up contracts. The AAA, however, instructed landlords to divide the plow-up payments with their tenants according to the interest each renter and cropper held in the cotton crop. A sharecropper was to receive half and a share renter was to receive two thirds or three fourths of the plow-up payments. In some cases, landlords followed these recommendations and divided the payments with their tenants. More commonly, landlords used their tenants' shares of the plow-up payments to pay off the tenants' old debts. Some unscrupulous planters even claimed that they had no tenants, and they pocketed all the plow-up payments.[53] A sociologist, who investigated the 1933 "plow-up" program, tersely concluded that "a considerable number of tenants did not receive the full amount specified by the contract."[54]

Tenants fared even worse under the AAA cotton-control program of 1934–1935. Oscar Johnston, the manager of the D & PL estate in the Mississippi Delta (see chapter 4), was also the financial director of the AAA, and helped plan the new program that favored the landowning cotton growers. Each participating landlord was to receive an acreage allotment based on 40 percent of his cotton production during the years between 1928 and 1932. In return for removing this land from cotton cultivation, the landlords received "rental payments" of 3.5 cents per pound for previous acreage yields on acres withdrawn from production as well as "parity payments" of 1 cent for each pound formerly grown. The rental payments went to the landlords, who had to offer half to the share renters but none to the sharecroppers. The best a cropper could hope for was half of the parity payment or only half-a-penny per pound on cotton formerly grown. Sharecroppers soon called their pittance the "poverty payment."[55]

Although the AAA cotton-control program offered little money to croppers, Section VII of the 1934–1935 contract was supposed to protect the rights of tenants. According to this section, landlords had to maintain the usual number of tenants on their estates, and they had to allow tenants to live rent free in their houses for two years. Despite Section VII, many landlords were tempted to evict some of their tenants because 40 percent of the cotton land now lay idle. Not surprisingly, a wave of evictions occurred during the 1934 season, as landlords disposed of surplus tenant families.[56]

Not only did some landlords violate AAA policy by evicting tenants, but they also tried to seize their tenants' AAA payments. Section X of the 1934–1935 cotton-control program required landlords to share parity payments with tenants and rental payments with share renters. Yet, some landlords demoted share renters to sharecroppers so they could collect all the rental payments, and they reduced sharecroppers to wage hands so they could collect all the parity payments. Other landlords simply claimed the tenants' parity payments, applying the money to their tenants' outstanding debts.[57] Although some landlords did share the AAA payments with their tenants, the "withholding of tenants' parity payments by landlords, either in payment of debts or out-and-out cheating, was probably widespread."[58]

Because the AAA had to process over 1 million cotton-control contracts during 1934 and 1935, the agency found it difficult to prevent landlords from cheating and evicting tenants. One notorious example of landlord abuse, however, became widely known in December 1934, when two dozen tenants sued Hiram Norcross, an Arkansas landlord, in the Poinsett County court. Norcross, the owner of the Fairview Farms estate near Tyronza, had openly cheated and evicted tenants during the 1934 season.[59]

The tenants who sued Norcross were members of the Southern Tenant Farmers' Union (STFU), which had been founded by disgruntled tenants on Norcross' plantation in July 1934. Aided by Harry Mitchell and Henry East—two White socialists from Tyronza, the Black and White tenants banded together to protest abuses by Norcross and other landlords. Led by a White sharecropper president, J. R. Butler, and a Black sharecropper vice president, E. B. McKinney, the STFU became a biracial workers' union that threatened the local economic and social order.[60] As one STFU supporter noted: "The plantation owners have not succeeded in inculcating hatred between the whites and the Negroes [*sic*], in as much as both races are willing, even eager, to unite their strength in the union."[61]

Although the biracial STFU ultimately lost the case against Norcross in the state supreme court, the Union created a furor in eastern Arkansas. Local planters formed a Klan-like group called the "Nightriders" to harass STFU organizers and members. Vigilante bands of planters, overseers, and deputies threatened, beat, and sometimes killed STFU supporters. Planters not only resorted to vigilante violence, but they also evicted dozens of STFU members from their plantation homes.[62]

Despite this "Terror in Arkansas," the Union survived the winter of 1935. After sending delegations to Washington, D.C. to protest the treatment of tenants, the Union organized a successful strike during the fall of 1935 to improve the wages of cotton pickers on Arkansas plantations. Thanks to such efforts, the STFU claimed more than twenty thousand members, and the Union had locals in six Southern states. Although angry planters continued to harass STFU members and supporters, the plight of cotton tenants was well publicized in the printed media. The troubles in eastern Arkansas dramatized the struggles of tenants throughout the Southern bottomlands.[63]

Although serious questions about the treatment of cotton tenants dogged the AAA crop-control programs, these "New Deal" efforts did have a positive effect on the Southern cotton economy. The AAA programs reduced the huge cotton crops that had glutted the world market. During the 1933 crop year, the United States produced 13,050,000 cotton bales. The following year, the crop fell to only 9,640,000 bales. And in 1935, the United States harvested 10,640,000 bales. These smaller crops, in turn, helped push up cotton prices. Between 1933 and 1935, the farm price of cotton lint averaged more than eleven cents per pound.[64]

Encouraged by efforts to control crops between 1933 and 1935, the AAA Cotton Division, as the cotton section was now known, launched its sign-up campaign for the 1936–1939 cotton-control program. Yet, on January 6, 1936, the U.S. Supreme Court invalidated the AAA of 1933, because of the controversial processing tax on commodities. Congress responded quickly to the court's attack on the crop-control program. In February, Congress passed the Soil Conservation and Domestic Allotment Act. Although this act provided crop controls and benefit payments for the purpose of "soil conservation," it retained the goal of curbing production in order to raise commodity prices.[65]

The act, which superseded the first AAA, proved less successful in curbing crop surpluses. The new act encouraged growers to reduce the acreage of such soil-depleting crops as cotton and to increase the acreage of such soil-building crops as legumes. To reward cooperating growers, the U.S. Department of Agriculture used a yearly appropriation to pay cooperators about ten dollars an acre for taking land out of cotton production. Although this program succeeded in fostering soil conservation, it did not succeed in controlling cotton production. During the 1936 crop year, U.S. growers produced 12,400,000 cotton bales, and they harvested a banner crop of 18,950,000 bales during 1937. As crops increased in size, the farm price of cotton decreased from twelve cents a pound in 1936 to little more than eight cents a pound the following year.[66]

Given the failure of the Soil Conservation Act to control crop production, Congress passed a second AAA in February 1938. This new act emphasized crop-control rather than soil conservation. Each year, the U.S. Department of Agriculture would estimate the cotton acreage needed to meet domestic and foreign demands. This allowable acreage was then divided among the cotton states. Within each state, elected county committees determined each grower's share of the county's cotton allotment. If growers cooperated with this second AAA program, they received payments for reducing cotton acreage, payments for planting soil-conserving crops, and even parity payments. By 1939, about 95 percent of Southern cotton growers were participating in the second AAA crop-control program. U.S. cotton acreage fell from 33 million in 1937 to about 24 million in 1938 and 1939. Despite the acreage reduction, the huge cotton crop of 1937 glutted the market, and cotton prices remained sluggish. Between 1937 and 1940, the price of cotton averaged less than nine cents per pound.[67]

In retrospect, the New Deal crop-control programs achieved a rather mixed economic record. The first AAA curbed cotton production, and cotton prices rose between 1933 and 1936. The Soil Conservation and Domestic Allotment Act, however, failed to limit cotton production, and prices dropped in 1937. Although the second AAA curbed production after 1938, cotton prices stagnated during the late 1930s. Not only were the economic results mixed, but the social benefits of these "New Deal" crop-control programs were also mixed. Landowning growers generally benefited from "New Deal" programs, whereas landless growers often failed to share in the rewards.

After 1933, landowning cotton growers, who participated in the "New Deal" programs, received an array of plow-up, crop-control, soil conservation, and parity payments. These federal payments pumped much needed cash into the Southern cotton economy. Thanks to government benefit payments, many struggling landlords reversed their financial fortunes during the "New Deal" (1933–1941) years.[68] Thad Snow, a Missouri Delta farmer, fondly recalled the "New Deal" era, when "the farmers of the Delta, or at any rate, the landowning farmers, got accustomed to making money, and they expected to make more and more money in the years ahead."[69]

The leading recipient of federal benefit payments was the British-owned D & PL estate in the Mississippi Delta—the largest cotton plantation in the United States. Managed by Johnston, who also worked for the AAA, the D & PL estate reversed its fortunes during the 1930s. In the years between 1930 and 1932, when cotton prices averaged only seven cents a pound, the D & PL estate ran deficits averaging $198,000 a year. But during the years between 1933 and 1935, the estate earned operating profits averaging $250,000 a year. This reversal of fortunes was partly due to rising cotton prices, which averaged eleven cents per pound during these years, and it was partly due to the AAA cotton-control program. Between 1933 and 1935, the D & PL corporation received over $318,000 in federal benefit payments.[70]

Skillfully managing the huge D & PL estate, Johnston claimed that the future of cotton planting lay with corporate plantations that treated workers in a paternalistic fashion.[71] Johnston firmly believed that a sharecropper who was considerately treated worked better than one who was abused and exploited.[72] Unlike some plantations, which cheated tenants at "settling time," the croppers on the D & PL estate received fair settlements. They even collected their fair share of the federal benefit payments. Finally, the D & PL croppers generally earned more money than most Southern cotton tenants.[73] As Johnston liked to say to his croppers: "When you make money, I make money. When you don't, I don't; and when I don't, you don't."[74]

Pursuing a paternalistic policy, the D & PL corporation enjoyed its most profitable year in 1936. In that year, the estate planted about 10,000 acres in cotton and 7,200 acres in other crops. Receiving $68,000 in federal soil conservation payments, the D & PL estate grossed $879,000. After deducting $361,000 in expenses, the estate earned an operating profit of $518,000.

Not only did the D & PL corporation earn an impressive operating profit, but the hundreds of Black croppers who worked on the estate also fared rather well in 1936. In that year, the average D & PL cropper family received $205 in credit at the plantation store during the growing season, and they earned $320 in cash at the end of the season. Many Southern cotton tenants were fortunate to earn as much as one hundred dollars in cash at the end of the season.

One of the most successful D & PL croppers was forty-eight-year-old Lonnie Fair. The head of a family of seven, Fair cultivated twenty-two acres of cotton and five acres of corn with the help of his family and the use of D & PL mules. In addition to furnishing mules, the D & PL estate also provided Fair with a house, garden plot, woodlot, cottonseed, and farming tools. Although Fair charged $483 worth of goods at the plantation store, he cleared $518 in cash, when he sold his share of the cotton. Earning a sizable cash income, the Fairs also raised a vegetable garden, kept two milk cows, and fed four pigs. Fair's success was not an accident because he had been a cropper on the D & PL estate for fifteen years.[75]

By treating croppers more fairly than most planters, the paternalistic manager of the D & PL estate earned the loyalty of his workers. In 1939, the D & PL estate housed 821 cropper families, and the average tenure of these croppers was thirteen years. Unlike most Southern cotton plantations, where many of the tenants left at the end of the year, Johnston retained a stable core of skilled croppers over the years.

By 1939, the D & PL estate was participating in the second AAA cotton-control program. The estate planted only eighty-five hundred acres in cotton that year, but it cultivated eight thousand acres in forage crops. Because cotton acreage was reduced to ten acres per cropper family, they earned an average of $152 in cash from their share of the cotton crop. In addition, each cropper received about $100 in crop-control payments, and they also worked as part-time day workers on the plantation's forage crops.[76]

Although the D & PL croppers fared better than most bottomland tenants, and although the D & PL estate earned higher profits than most bottomland plantations, Johnston's example of paternalistic capitalism was not widely emulated. Despite its financial success, the D & PL estate was something of an anachronism. Johnston sparingly used tractors and machines on the D & PL plantations: He preferred to work the estate with skilled croppers and trained mules.[77]

A Southern journalist, who visited the D & PL estate, admired Johnston's efficient operation, which relied on men and mules rather than tractors and machines:

> Here on the Mississippi Delta, Johnston seemed to me to be leading not so much a resistance against the machine as a demonstration that mules and men, well-cared for and well-fed, under wise direction, even in the old plantation pattern, might still together live well and happily on the rich earth of the warm South. For a little while at least.[78]

Although Johnston was reluctant to mechanize his operations, some bottomland planters adopted "power-drive agricultural implements," and they hired "day labor or wage hands" to do "the work that cannot be done mechanically, like chopping and picking."[79] Thanks to "New Deal" programs, bottomland planters obtained the capital and credit needed to mechanize their operations. Federal benefit payments supplemented plantation incomes, and federal lending agencies extended low-interest loans to planters.

After 1933, federal benefit payments became an essential part of plantation income. In 1934, when an economist surveyed eighty-nine plantations in nine bottomland Arkansas counties, he found the estates earned an average of $10,774 in gross income. After deducting $6,271 in expenses, these plantations earned an average of $4,503 in net cash income. Government benefit payments accounted for $1,935 or 43 percent of the income.[80]

Bottomland planters not only collected government benefit payments after 1933, but they also found it easier to secure credit. Planters needed long-term and short-term credit in order to operate their estates. Long-term loans were needed for capital investments in land, machinery, livestock, and buildings. Short-term loans, in turn, were required to plant crops, furnish tenants, feed mules, operate tractors, and pay wages.[81] In his survey of Arkansas bottomland plantations in 1934, economist H. W. Blalock found that a typical thirteen-hundred-acre plantation had $71,224 worth of capital invested in lands, machines, animals, and buildings. In addition to these long-term capital investments, a typical Arkansas plantation also needed more than four thousand dollars in "operating capital" each season "for making the crop."[82] To obtain long-term capital, planters mortgaged their estates, borrowing money from commercial banks, insurance companies, and the federal land banks. The federal land banks, which were created by the Farm Loan Act of 1916, made substantial loans that were secured by first mortgages on estates. Offering long-term loans at low interest rates, the farm land banks were expanded by the Farm Mortgage Acts of 1933. In order to obtain short-term operating capital, planters had to mortgage machinery, livestock, and forthcoming crops, borrowing the money from banks, merchants, and federal credit associations. In 1933, Congress passed the Farm Credit Act, which established production credit associations. These credit associations offered short-term loans for three to twelve months, which were secured by crop liens and chattel mortgages. Production credit associations proved a boon to cotton planters, furnishing a steady and secure source of operating capital at low interest rates.[83] Thanks to the availability of federal credit association loans, federal land bank loans, and government benefit payments, cash and credit were available for investment in land, machinery, livestock, and wage labor.

During the 1920s and the early 1930s, it was "the rule that the more land that a man owned and operated the sooner and surer he'd go broke." But after 1933, the "more land a man bought, even on credit, and the more he operated, the sooner he was sure to get rich." The insurance companies and other lending institutions, which had foreclosed on the lands of many bankrupt planters and farmers during the depression, were anxious to dispose of their devalued landholdings. The insurance companies sold lands to "hustling, acquisitive operators," who would make a modest down payment and assume a debt that approximated the amount of the original mortgage of the previous owner. Given the improving economy after 1933, these operators often paid off their debts in a few years. In these unsettled times, one never knew "who might blossom out into a really big operator."[84]

One such hustling operator was Johnie [*sic*] Brown, a Black farmer from Issaquena County, Mississippi, who entered the ranks of landowning planters in 1934. An insurance company, which held the title to an abandoned 580-acre plantation, sold the estate to Brown for only seven dollars an acre. Requiring no cash down, the company gave Brown ten years to pay for the land. Paying for this plantation long before the ten years was up, Brown "just kept plugging." He invested every dollar of profit in land until he accumulated thirty-seven hundred acres.[85]

Acquiring large tracts of land, planters needed "power machinery" for their "large-scale operations."[86] Using credit and cash, the planters in Coahoma County, Mississippi, spent more than $250,000 on tractors in 1934–1935. A single planter in Washington County, Mississippi, owned 22 tractors and 13 four-row cultivators, which tilled as much cotton as 130 tenants. By 1939, Washington County counted 846 tractors, the highest number in any Mississippi county, and Coahoma claimed another 541 tractors.[87]

By substituting tractor power for mule power, it was possible to raise more cotton on more land at less cost. In 1931, an economist compared the records of sixty-six Mississippi Delta tractor-powered plantations with twenty-eight mule-powered estates. The tractor-using estates employed general purpose (Farmall) tractors to pull disk plows for land-breaking, two-row middle-breakers for making cotton beds, four-row planters to seed the beds, and four-row cultivators for doing the preliminary cultivation. In turn, the mule-powered plantations used mule teams to pull disk plows, middle-breakers, planters, and cultivators. On the average, it required 6.2 tractor hours and cost $3.80 per acre to plow, bed, plant and cultivate an acre of cotton with tractor-drawn equipment, whereas it required 43.1 mule hours and cost $5.91 per acre to do this work with mule-drawn equipment.[88]

Tractors replaced the mule power needed to grow cotton, but these machines did not replace the human labor needed to hoe, chop, and pick cotton. Using hoe hands, cotton pickers and one-row mule-drawn tools, it required 150 man hours to produce an acre of cotton on a Mississippi Delta estate. Using tractor-drawn tools to plow, plant, and cultivate cotton, it still required 136 man hours to produce one acre of cotton.[89] Despite partial mechanization, tractor-using planters still needed numerous laborers to hoe, chop, and pick cotton. In the 1931 study of Mississippi Delta plantations, economist Lewis Long found that tractor-using planters relied on four types of hired laborers to cultivate and harvest cotton. The best laborers were Black sharecroppers, whom the planters employed when the croppers were "not busy with their own crop." Although these hired croppers often proved adequate for hoeing and chopping cotton, planters needed additional laborers for cotton picking. Tractor-using planters generally employed temporary Black laborers from neighboring towns and cities, hauling these day hands in trucks to their plantations. Less commonly, planters imported temporary Black workers from outside the Mississippi Delta. Tractor-using planters had to recruit and transport these Black laborers to their estates, but the workers had to return "to their homes at their own expense." Because these temporary Black workers could leave at any time, they were not as dependable a source of labor as the Mexican migrant workers from southern Texas. Mexican workers gave "good service" and stayed until the end of the harvest, but they had to be transported to and from their homes, and these transportation expenses added to the cost of cotton picking. Proving expensive and often undependable, hired cotton pickers did not solve the problem of harvesting cotton. Cotton harvesting remained a persistent problem that limited "the production of cotton with tractor power."[90]

Because cotton growing still required a great deal of human labor, it was impossible to fully mechanize cotton production in the 1930s. Other crops, however, proved more amenable to mechanization. It was possible to sow, tend, and harvest oats, soybeans, alfalfa, and lespedeza with tractor-drawn equipment (see chapter 4). Because the government crop-control programs had removed much of the cotton acreage from production, planters were free to sow these idle acres with tractor-raised grains and legumes.[91]

The federal cotton-control programs thus initiated a revolution in land-use practices during the 1930s. An economist, who studied plantations and farms in southeastern Arkansas and northeastern Louisiana, found striking changes in the crop acreage between 1929 and 1939. In six southeastern Arkansas and in six northeastern Louisiana parishes, cotton acreage declined from 901,295 acres in 1929 to 595,589 acres by 1939. In contrast, the acreage devoted to grains, legumes, and grasses soared from 322,043 acres in 1929 to 811,263 acres by 1939. As the acreage in pastures, meadows, and forage crops increased, there was a related growth in cattle and hog populations in southeastern Arkansas and northeastern Louisiana. The number of cattle grew from 128,056 in 1929 to 226,103 by 1939, and the hog population increased from 95,648 to 174,022. Because livestock and feed crops were replacing cotton, and because livestock and feed crops required less labor than cotton, there was less need for tenant farmers to raise and pick cotton.[92]

An economic study of land tenure in Arkansas, which was published in 1940, documented the decline of tenant farmers on bottomland plantations during the 1930s. Between 1932 and 1938, the number of share renter and share cropper families per ten thousand acres of crop land in three bottomland counties dropped from 941 to 689. Such "losses in the number of share-renter and share-cropper families were offset, to the extent that they were not displaced by mechanized equipment, by the increase in number of wage laborers." In the years between 1932 and 1938, the number of tractors per ten thousand acres of cropland in these bottomland counties grew from 40 to 71, and the number of wage hand families per ten thousand acres of cropland rose from 247 to 294. These dry statistics revealed that many tenants were evicted or demoted in eastern Arkansas during the 1930s.

Some tenant families were evicted from their plantation homes, when planters substituted tractor equipment and wage hands for mules and tenants. Those tenants, who did remain on plantations, were often demoted to a lower economic status. From 1932 to 1938, the "average cotton acreage per share-renter declined from 25 to 13 acres" in these bottomland counties, while "those of share-croppers declined from 15 to 10 acres." To make up for this loss in cotton acreage, the "share-renters and share-croppers relied more than formerly upon wage work as a source of income." In other words, cotton tenants were becoming tenant-wage hands.[93]

The demotion of tenants to tenant-wage hands began during the AAA crop-control programs of 1933–1935. Writing in 1935, a supporter of the Southern Tenant Farmers' Union described this process on a fourteen thousand-acre Arkansas plantation, which had "long been notorious for its bad treatment of its tenants":

> In the spring of 1934 the owners decided to change a part of this farm over to a day-labor basis. Croppers were notified that each adult worker could retain only 3.5 acres on a share-crop basis. In some cases this represented a cut of 75 per cent in acreage. They were required to cultivate a large part of the plantation on a day-labor basis at 75 cents for a thirteen-hour day.[94]

The replacement of tenants with tenant-wage hands became a common solution to the labor problem on many bottomland plantations. Because wage workers were rather undependable, planters created a more dependable labor force by devising "a cropper-wage type of arrangement." A "cropper family was given a small share-crop, inadequate to provide sufficient income for the year, but sufficient to induce the family to remain located on a given farm and to see the crop through." For "the remainder of its income, the family was dependent on wage employment which the provider of the share-crop could offer at his option." These cropper-wage hands became a dependable band of workers for hoeing, chopping, and picking cotton. Planters augmented these cropper-hands with temporary wage workers, especially during the cotton-picking season.[95]

Relying on wage workers, cropper-hands, and tractor-drawn equipment, bottomland planters reduced the cotton acreage, increased the acreage of forage crops, and raised livestock as a supplementary "cash crop." By 1939, the revenue from the sale of beef cattle in Arkansas rivaled that from the sale of cotton.[96] As cotton fields gave way to livestock pastures and fields of forage crops, the "fragmented" plantations, which had dominated the bottomland landscape since the end of the Civil War, began evolving into newly consolidated plantations.[97]

For decades, planters had operated "tenant" plantations, which had been subdivided into "tenant tracts" and "home farms." The bulk of the crop land was divided into small tenant tracts, which were worked by croppers and renters. The remaining crop land became the "home farms," which planters cultivated with hired workers (see chapters 1–4). But in the 1930s, the home farms began growing at the expense of the tenant tracts. Because the government paid planters to reduce cotton acreage, they planted this former cotton land, known colloquially as the "rented acreage," in forage crops, which they worked with tractor-drawn equipment and hired laborers.[98] Although the remaining cotton acreage was still worked by share croppers and cropper-wage hands, many planters adopted the "through-and-through" system. Tractors plowed and bedded the tenant acreage, which was then laid off and divided into cropper tracts. On some plantations, the croppers and cropper-hands simply tended "the common fields under a gang system directed by an overseer" [riding boss]. Consolidating tenant tracts into common fields, the new plantations proved more efficient than the fragmented plantations of earlier times.[99]

Although the transformation of fragmented plantations into consolidated estates improved efficiency, it exacted a human toll. As planters shifted from mules and tenants to tractors and hands, they evicted surplus tenant families. Evicted from their homes, displaced tenants became wage workers on plantations, they became migrant laborers who drifted from job to job, or they moved to the cities where they hoped to survive on relief.[100]

Because many displaced tenants ended up on the relief rolls, they became the responsibility of another "New Deal" agency—the Federal Emergency Relief Administration (FERA), which provided federal grants to states for use in direct relief. Administered by Harry Hopkins, the FERA planned to supplement direct relief with "rural rehabilitation" loans. In April 1934, Hopkins created the Division of Rural Rehabilitation, which loaned money to struggling farm families so they could stay on the land. To supplement this rural rehabilitation program, William R. Dyess, Arkansas' FERA administrator, devised another way of helping the rural poor. He planned to "resettle" deserving relief clients in government-sponsored agricultural communities.[101]

In early 1934, Dyess used federal funds to buy more than fifteen thousand acres of cut-over timberland in Mississippi County, Arkansas. Launching his "Colonization Project No. 1" in May, Dyess' crew laid out five hundred farmsteads as well as a community center. Recruiting relief clients from across the state, Dyess began resettling these families on farmsteads ranging from twenty to forty acres in size. Although Dyess, who was killed in an airplane crash in January 1936, failed to see the dedication of his project on May 22, 1936, the Dyess resettlement colony was a bold experiment that inspired other resettlement communities in the Southern bottomlands.[102]

Among the relief clients who settled at the Dyess colony was Ray Cash, his wife, and their six children. One of the children was Johnny Cash, who later achieved fame as a country musician. Johnny was almost four years old, when his family arrived at the colony in the winter of 1935. Years later,

Johnny remembered the ice hanging off the trees, as his family drove through the colony in their old truck. Arriving at House No. 266, they moved into a new dwelling on twenty acres of timberland. Ray and his eldest son hacked out ten acres of farmland during the first year. Although they had to battle numerous water moccasins, the Cash family managed to plant cotton on their hard-won land, gathering two bales per acre. Succeeding on their farm, the Cashes regarded the Dyess colony as their "salvation."[103]

Work was continuing on the Dyess colony, when the Roosevelt administration created the Works Progress Administration (WPA) and the Resettlement Administration (RA) in the summer of 1935. Superseding the FERA, the WPA, which was directed by Harry Hopkins, planned a host of public works projects to provide "work relief" for urban and rural poor. In turn, the RA, which was led by Rexford Tugwell, assumed responsibility for rural resettlement and rural rehabilitation. The RA contained a Rural Rehabilitation Division, which could make rehabilitation loans to farm families, as well as a Resettlement Division, which would resettle poor people on the land. In order to implement its programs, the RA established a dozen regional offices across the United States, including the Region VI office in Little Rock, Arkansas. Directed by T. Roy Reid, the RA's Region VI office managed rural rehabilitation and rural resettlement in Arkansas, Louisiana, and Mississippi. In addition to making rehabilitation loans to farm families, the Region VI staff planned resettlement projects to shelter needy farm families. The Dyess colony was not among these RA projects because the colony remained under WPA control for another three years, but the Dyess project served as a model for the RA resettlement efforts in the Southern bottomlands.[104]

These RA projects, however, did little to calm the turmoil among Arkansas' tenant farmers. In January 1936, the STFU set up a tent colony near Parkin, Arkansas, to dramatize the plight of evicted tenant families. A few months later, the STFU organized a cotton choppers' strike in eastern Arkansas to protest poverty among tenant farmers and wage hands. During this strike, local bullies beat two White socialists—Rev. Claude Williams and Willie Sue Blagden—who were helping the STFU, and these beatings received national attention. Responding to the negative publicity, Governor Marion Futrell of Arkansas appointed a state commission in August 1936, to investigate the tenancy problem and to recommend solutions. This Arkansas Tenancy Commission, which included STFU supporters, submitted its report in November, condemning the institution of tenancy and recommending a federal farm purchase act to help tenants become landowners.[105]

Tenancy not only became a political issue in Arkansas, but it also surfaced in national politics. After Franklin Roosevelt was re-elected in November 1936, he created a presidential Committee on Farm Tenancy, which was chaired by Henry Wallace, the secretary of agriculture. Meeting in December, the committee counted thirty-three leading citizens, including Arkansas' new governor—Carl Bailey. Completing its report in February 1937, the committee recommended a "Farm Security Administration" with a tenant purchase program to enable tenants to buy farms on easy terms. Following the committee's report, Sen. John Bankhead of Alabama and Rep. Marvin Jones of Texas introduced bills that put the committee's recommendations into effect. Passed by Congress, the Bankhead–Jones Farm Tenant Act became law in July 1937. The act provided for long-term, low-interest loans to tenant farmers, which would enable them to buy their own land. Following the recommendations of the presidential committee, Roosevelt abolished the old RA in September, and he created the new Farm Security Administration (FSA), which was directed by Will Alexander. The FSA not only inherited the RA's rural rehabilitation and rural resettlement programs, but it also administered the new tenant purchase loan program.[106]

Because the old RA bureaucracy was largely absorbed by the new FSA, the Region VI office remained under T. Roy Reid, who added a Tenant Purchase Division to supplement the Rehabilitation and Resettlement Divisions. Although Region VI had many dedicated officials, the FSA was underfunded by Congress. As a result, the FSA helped only a fraction of the needy tenant farmers in Region VI. In the FSA's Region VI, about 146,000 farm families received rural rehabilitation loans, another 7,000 obtained tenant purchase loans, and the resettlement projects sheltered nearly 3,000 families.[107]

Due to limited resources, the FSA was unable to help the mass of poverty-stricken tenants—many of whom were Black. FSA rural rehabilitation loans usually went to White farmers, and FSA tenant purchase loans generally went to White tenants with "good credit ratings." Even the FSA rural resettlement program favored the more successful White tenants because the agency wanted to make a "good showing" with its projects. Of the thirty resettlement projects in Region VI, which included the Dyess colony after 1939, twenty-five were all White or predominantly White, and only five were all Black. These Black resettlement projects included Desha, Lakeview, and Townes farms in eastern Arkansas, Mounds Farms in northeastern Louisiana, and Mileston Farms in the Mississippi Delta. In addition, some two hundred Black families lived in predominantly White projects within segregated areas.[108]

In contrast to the segregated government resettlement projects, the Delta Cooperative Farm near Hill house, Mississippi, was a "joint enterprise of dispossessed white and Negro [*sic*] sharecroppers." It was "created partly as a place of refuge from the terrorism that swept the nearby cotton counties of eastern Arkansas in 1934 and '35, and under which no sharecropper's life was secure." The Hillhouse Farm was a branch of the STFU, and had the support of such Christian philanthropists as Dr. Sherwood Eddy.[109] Borrowing funds from Eddy and other benefactors, the Delta Cooperative purchased 2,133 acres of "buckshot" clay land for six dollars an acre. Only 350 acres of this land had been cleared for farming. Planting 350 acres of cotton in 1936, the cooperative sold the lint and seed as well as some cypress timber for $18,000. After deducting the operating expenses and the interest on the debt, this left a net profit of $8,900. Divided among the thirty White and Black families in the cooperative, the profit yielded an average cash income of around three hundred dollars per family. This was a larger cash income than most sharecropper families earned in the cotton South in 1936. In addition to this cash income, the cooperative members shared the produce from a ten-acre vegetable garden and a communal poultry farm. Each family in the cooperative also had the use of a frame cabin with screened porches. There were two rows of cabins—one for Black families and one for White families—but all the families shared a community center that contained a library, clinic, nursery, and meeting house. White and Black families attended the integrated cooperative meetings, and there were two Black members on the five-person farm council.[110] This "social act of strict heroism" was being "conducted in the face of sneers and threats from neighboring planters and riding bosses." The Hillhouse Farm remained "a small ship afloat on a vast, angry sea," but the farm demonstrated "that white and Negro [*sic*] sharecroppers, organized cooperatively," could "raise their standard of living far above anything the South has previously known."[111] Although the Hillhouse Farm was an isolated example, it showed how private philanthropists could have aided the tenant families of the Southern bottomlands.[112]

Because the Delta Cooperative Farm and the federal resettlement projects "resettled" only a few thousand families in the Southern bottomlands, these token efforts had failed to shelter the majority of tenant families who were evicted from plantations.[113] One economic historian estimated that one fifth of the tenant labor force in the bottomland counties of Mississippi, Louisiana, and Arkansas had been evicted during the 1930s.[114]

The STFU had protested these tenant evictions since its inception in 1934. STFU activities had focused national attention on the plight of displaced tenants, and this attention had contributed to the creation of the FSA.[115] But by 1937, the STFU was in disarray. In January, the STFU affiliated with the Congress of Industrial Organization (CIO) as part of the United Cannery, Agricultural, Packing & Allied Workers of America (UCAP & AWA). UCAP & AWA was led by Donald Henderson, an avowed communist, a fact that touched off an internal struggle within the socialist-led STFU. When STFU Vice President E. B. McKinney sided with Henderson, he was expelled by STFU President J. R. Butler and STFU Secretary Harry Mitchell. They replaced McKinney with the Rev. Owen Whitfield—a Black preacher, who was an STFU organizer in southeastern Missouri.[116]

Whitfield had been a share cropper on a Missouri Delta farm. When he was evicted for STFU activities, a friendly White farmer, Thad Snow, helped him secure a homestead on the FSA LaForge resettlement project near Sikeston, Missouri. With Snow's help, Whitfield continued to spread the message about the STFU. Although Snow was a landowning farmer, he was disturbed by the conduct of his fellow landlords. Many were demoting croppers to laborers so they could collect the tenants' share of the government benefits, and they were evicting unwanted croppers and laborers at the end of the cotton season when they were no longer needed. A wave of evictions in late 1938 inspired Whitfield to make a dramatic protest.[117] He planned a roadside "sit-down strike." On the morning of January 10, 1939, hundreds of evicted croppers and laborers camped along the major highways in southeastern Missouri in order to protest their treatment on local farms. This novel roadside protest soon attracted national publicity. Reporters from newspapers and magazines flocked to southeastern Missouri to interview and photograph the cold, hungry protesters. Angered by the negative publicity, the state authorities stifled the protest. On January 15, the state health commissioner condemned the roadside camps as a threat to public health. The next day, state policemen cleared the camps, loaded the protesters into trucks, and scattered them across southeastern Missouri. Whitfield was not among these protesters. After organizing the roadside protest, he escaped to St. Louis in order to avoid a lynching. During his exile in St. Louis, a Sikeston newspaper editor accused Snow, Whitfield's friend, of being the "brains" behind the roadside protest. Following this accusation came talk of shooting Snow. Fortunately, Snow deflated the hysteria by publishing a satirical "confession" in a local newspaper, claiming that he had hatched the roadside protest plot in collusion with Leon Trotsky. During these dangerous weeks, Whitfield remained in St. Louis, where he organized a Committee for the Rehabilitation of the Sharecroppers. Because Whitfield had organized the roadside protest without the knowledge of the STFU leaders, and because he was sympathetic to the UCAP & AWA leaders, Whitfield lost his job as STFU vice president. In March, after suspending Whitfield, Butler and other STFU socialists seceded from UCAP & AWA. The break with UCAP & AWA left the STFU with fewer than forty locals—most of them located in northeastern Arkansas. Whitfield, McKinney, and many of the Black STFU members remained in the UCAP & AWA.[118]

Although Whitfield's roadside protest contributed to the breakup of the STFU, he focused national attention on the desperate tenants and laborers of southeastern Missouri. In spring 1939, Whitfield and his wife, Zella, traveled to Washington, D.C., where he testified before a Congressional committee and his wife met with Eleanor Roosevelt. The federal government responded to Whitfield's protest by redoubling the efforts of the FSA in southeastern Missouri. Earnest FSA bureaucrats arrived from the Indianapolis, Indiana, regional office to award cash grants, make rehabilitation loans, and establish resettlement communities such as the FSA Delmo Labor Homes. These new FSA efforts far surpassed the token programs of the agency in southeastern Missouri before the roadside protest.[119]

Despite the FSA efforts in southeastern Missouri and elsewhere, landlords continued to mechanize their estates, hire wage workers, and evict unwanted tenant families. Because the evolving cotton plantations displaced so many tenants, and because the cotton plantations underwent such striking changes, bottomland cotton planting has received far more attention from historians than the Arkansas rice industry.

After surviving the agricultural recession of the 1920s, Arkansas' rice farmers faced a new threat during the Great Depression. In 1929, the farm price for rice was ninety-seven cents a bushel. Three years later, the price fell to forty-seven cents.[120] By 1932, Arkansans were sowing only 163,000 acres of rice. Facing the lowest prices since the Civil War, rice growers welcomed the creation of the AAA, which included a rice section. Unfortunately, the AAA rice section proved far less effective than the AAA cotton section. After months of debating, the rice section finally unveiled a crop-control program in July 1934. Arkansas rice growers were to cut their production by one fifth, using the past five years' average as the base period. To reward cooperating growers for cutting production, the rice section devised a cumbersome "60–40 plan." Rice millers paid growers 60 percent of the "parity price" at the time of purchase, and the remaining 40 percent went into a AAA trust fund that would be disbursed to cooperating growers. Unlike the AAA "cotton section," which favored landlords over tenants, the rice section offered allotments to both land-downing landless rice growers. But when rice millers refused to cooperate, the rice section abandoned the unworkable 60–40 plan at the end of the year. In 1935, the rice section drafted another crop-control program. Rice millers paid a processing tax of one cent per pound on rough rice, and this processing tax in turn paid the cooperating growers who cut rice production by one fifth. Rice millers were still protesting this processing tax, when the Supreme Court invalidated the first Agricultural Adjustment Act in January 1936. Congress quickly passed the Soil Conservation and Domestic Allotment Act, which offered inducements to rice growers who reduced acreage. The improving economy, however, buoyed up crop prices, and growers increased their rice acreage. Two years later, Congress passed the second Agricultural Adjustment Act that provided marketing quotas when the rice supply exceeded "normal supply" by more than one tenth. Despite these federal efforts to control rice production, rice acreage continued to increase. By 1940, Arkansans were harvesting 191,000 acres of rice. Thus, despite "confusion and poor administration in the AAA rice program," Arkansas' rice industry weathered the 1930s "with few changes."[121]

Although the struggles of rice farmers, cotton planters, and cotton tenants have dominated the historiography of the Southern bottomlands during the Great Depression, there was another story that has often been overlooked. Despite the Great Depression, the Southern bottomlands continued to attract settlers. These Depression-era settlers ranged from unemployed urbanites to landless farmers to land seekers.

During the agricultural recession of the 1920s, many Black and White farmers had left the rural South, seeking a better life in the industrial cities. Yet, when the Great Depression began, many of these urban workers lost their jobs, and during the early 1930s, these unemployed urbanites drifted back to the countryside. Thousands of these White and Black urban refugees sojourned in the Southern bottomlands, renting tenant tracts or squatting on public lands.[122]

In addition to these urban refugees, landless farmers sought refuge in the Southern bottomlands. Fleeing boll weevils as well as the Great Depression, many Southern cotton-tenant families sought refuge in the Missouri Delta, where weevils posed little threat. Although some of these Southern refugees managed to purchase devalued farmland, most rented or cropped corporate acreage. The major insurance companies had foreclosed on many indebted farmers during the early 1930s, and these corporations had amassed thousands of acres of farm land. Among the largest corporate landholders was the General American Life Insurance Company, which had acquired the landed assets of the bankrupt Missouri State Life Insurance Company in 1933. The following year, General American Life employed Maxwell Williams to manage nine thousand acres of corporate farmland in Pemiscot County. In order to work this land, Williams rented acreage to tenants who owned mules and tools. These renters, in turn, furnished mules and tools to share cropper families who raised and picked the cotton. The share croppers received half of the cotton crop that they raised, the renters got one fourth of the crop, and the corporation received the remaining one fourth as rent. Relying on tenants and mules to raise cotton, the corporation built decent houses and barns to shelter families and work stock. Pursuing a paternalistic policy on its Pemiscot farmland, General American Life became one of the leading cotton producers in the United States.[123]

Although some tenant families found refuge on the corporate farms of southeastern Missouri, landless Southerners still dreamed of owning their own farmland. The dream of land ownership compelled more than twenty thousand families to settle in "the undeveloped areas" of the lower Mississippi River Valley during the 1930s, and most of these newcomers located on "small family-size tracts" in eastern Arkansas, northwestern Mississippi, and northeastern Louisiana. They were lured by "huge tracts of poorly-drained, cut-over lowland," which were in the hands of state governments, land speculators, and corporate owners. The bottomlands of Arkansas, Mississippi, and Louisiana contained "at least 5,000,000 acres of unimproved Delta land," which was suitable "for agricultural development."[124] Given the wealth of arable alluvial land, the bottomlands attracted farmers who were "tired of hillside farming," young couples who lacked "opportunities in the older areas," and "tenants [renters] and sharecroppers [who were] 'tractored off' the Delta plantations."[125]

In Arkansas, hundreds of hungry but ambitious farm families, "who desperately wanted to better their situation," left the eroded hill counties during the 1930s. Moving into the bottomlands, they found thousands of acres of tax-forfeited alluvial land. Buying this tax-forfeited land from the state for back taxes, they often paid as little as $250 for forty acres of fertile alluvial land.[126]

The fertile lands of the Mississippi Delta also lured land-hungry Black and White families. In the Delta, land speculators purchased tax-forfeited lands from the state government, and they sold this acreage to buyers, who included "natives of the Delta" as well as newcomers from the hill counties. Typically, families purchased low-lying, cut-over timberland, which was dotted with stumps and covered with brush and small trees. Purchasers not only had to clear and ditch this land, but they also had to construct shelters for their families and work stock. Moving onto this undeveloped acreage, families fashioned farm buildings, dug drainage ditches, and cleared the overgrown, cut-over timberland by slashing and burning the brush. In "every respect," the rough-hewn buildings, the minimal improvements, and the partially cleared fields with stumps "still standing amidst the cotton and corn" gave "the impression of frontier conditions." Yet, unlike the old frontier farmers of the nineteenth century, who acquired cheap public land from federal and state governments, these new frontiersmen of the 1930s had to pay premium prices for their land. Although land speculators bought tax-forfeited lands for less than five dollars an acre, they sold acreage to farm families for about twenty-five dollars an acre. Buying overpriced farms on time, families had to make large annual principal and interest payments, or they lost their farms to foreclosure. Fortunately, some of the land speculators treated land buyers with leniency. If they fell behind in their payments, they were allowed to continue working their land. In the Mississippi Delta, the "paternalistic pattern of the planter-tenant relationship" tended "to be retained even between land-seller and settler."[127]

In contrast to the paternalistic land speculators of the Mississippi Delta, the corporate land sellers in northeastern Louisiana showed less leniency to land buyers. In East Carroll, Madison, and Tensas parishes, timber companies held thousands of acres of cut-over timberland, which they sold to land-hungry farm families. The White and Black land seekers included tenants, who had been displaced from plantations, as well as hill farmers, who abandoned their eroded lands. Taking advantage of these desperate people, corporations sold cut-over land, which was appraised for less than eight dollars an acre to farm families for twenty-five dollars to forty dollars an acre. Because the average land purchase was forty-four acres, and because the average purchase price was $28 an acre, the typical family assumed a debt of $1,232 at 6 percent interest. For the first three years, the families "leased" their farms. They cleared the timberlands, made improvements, and raised crops, but they paid no money to the land sellers. After three years, families began to pay off the principal and interest in ten large annual payments. Most of the new farmers earned less than five hundred dollars a year, thus causing them to struggle to make the yearly payments. If they failed to keep up the payments, they lost their farms to foreclosure, and the companies resold the repossessed land for higher prices because the acreage had been "improved."[128] Although hundreds of families bought cut-over bottomlands in the parishes, and put "forth great effort," they were "harried by debts and the fear of foreclosure.[129] Despite the high rate of failure among these "new ground" farmers, the dream of land ownership endured. When one family was foreclosed, another moved in to buy the farm on time. Bottomland farm families were still girdling trees, burning brush, and planting cotton amidst the dead trees, when the United States entered World War II.[130]

World War II (1941–1945) accelerated the transformation of the Southern bottomlands. When the war began in December 1941, thousands of tenants and laborers began leaving the bottomland plantations in search of war work in the industrial cities. As the rural exodus continued, there was a labor shortage in the cotton fields. Facing a labor crisis at a time when cotton demand was soaring, planters tried to raise cotton with imported Mexican migrant workers and with Axis prisoners of war. They also raised wages in order to attract the remaining bottomland laborers. Wages for cotton picking soared from one dollar per one hundred pounds in 1940 to three dollars per one hundred pounds by 1944. Anxious to reduce wage costs, planters readily embraced the mechanical cotton picker. On October 2, 1944, the International Harvester company conducted a convincing field demonstration with eight mechanical cotton pickers on Hopson Plantation near Clarksdale, Mississippi. Each machine did the work of fifty human pickers. Although these early mechanical pickers were quite expensive, they allowed planters to drastically cut the wage costs in cotton production. Using mechanical cotton pickers, it cost $5.26 to pick a bale of cotton, whereas it cost $39.41 to pick a bale with human pickers.[131]

Even the D & PL estate, which had long resisted mechanization, found it necessary to acquire mechanical cotton pickers. During the late 1930s, the D & PL estate annually housed around 850 cropper families. But by the end of World War II, more than 300 of the D & PL cropper houses were empty. The cropper families were gone, and they were not coming back. Because the D & PL estate had lost more than one third of its cropper labor force, it had to "use more modern farm methods" to enable the remaining 525 cropper families to produce what 850 cropper families had formerly produced. By 1947, the D & PL estate had five mechanical cotton pickers. And by 1955, D & PL hoped "to have as many as 150 [mechanical cotton-pickers] as well as other equipment—tractors, plows, cultivators, flame weeders, in adequate quantity" to put the estate "on a maximum machine basis."[132]

For more than a century, cotton farming in the Southern bottomlands had been "symbolized by the figures of a man and flop-eared mule, abaft and before a plow, and by the stooped form of the cotton picker, his long picking sack dragging behind him like a great worm." Thousands of men, women, children, and mules had spent their lives working in bottomland cotton fields.[133] But after World War II, bottomland cotton farming was characterized by tractor drivers, tractors, tractor-drawn implements, and mechanical cotton pickers.[134]

Appendix: Tables

Table 1. Population Density of the Southern Bottomlands (1880)

Regions*	Total Population	Area (square miles)
Mississippi Delta	109, 598	6,040
Louisiana Delta	90, 686	4,760
Arkansas Delta	136, 845	10,060
Missouri Delta	52, 885	3,320
Totals	390, 014	24,180

*Includes all counties and parishes listed in the preface, except Humphreys County, Mississippi (created in 1918).

Source: U.S. Census Office, *Tenth Census (1880): Report on Cotton Production in the United States* (Washington, D. C.: Government Printing Office, 1884), Part 1: 145–150, 318–323, 513–517, 585–596.

Table 2. Population Density of the Southern Bottomlands (1930)

Regions*	Total Population	Area (square miles)
Mississippi Delta	418, 003	6,056
Louisiana Delta	153, 006	4,808
Arkansas Delta	479, 019	9,705
Missouri Delta	171, 472	3,285
Totals	1, 221, 500	23, 854

*Includes counties and parishes listed in the preface.

Sources: U.S. Census Bureau, *Fifteenth Census of the United States: 1930: Population* (Washington, D. C.: Government Printing Office, 1932), Vol. 3: Part 1: 176–182, 974–979, 1274–1280, 1328–1337; U.S. Census Bureau, *Fifteenth Census of the United States: 1930: Agriculture* (Washington, D.C.: Government Printing Office, 1932), Vol. 2: Part 1: 982–987; Part 2: 1050–1055, 1136–1141, 1214–1218.

Table 3. White and Black Populations of the Bottomlands (1880)

Regions*	White Population	Black Population	Others	Total
Mississippi Delta	20, 825	88, 712	61	109, 598
Louisiana Delta	15, 923	74, 734	29	90, 686
Arkansas Delta	68, 264	68, 497	84	136, 845
Missouri Delta	47, 832	5, 052	1	52, 885
Totals	152, 844	236, 995	175	390, 014

*Includes counties and parishes listed in the preface, except Humphreys County, Mississippi (created in 1918).

Source: U.S. Census Office, *Statistics of the Population of the United States at the Tenth Census* (Washington, D.C.: Government Printing Office, 1883), Vol. 1: 381, 393–394, 397–399.

Table 4. White and Black Populations of the Bottomlands (1930)

Regions*	White Population	Black Population	Others	Total
Mississippi Delta	109, 518	306, 813	1, 672	418, 003
Louisiana Delta	70, 520	82, 328	158	153, 006
Arkansas Delta	282, 102	196, 605	312	479, 019
Missouri Delta	148, 106	23, 338	28	171, 472
Totals	610, 246	609, 084	2, 170	1, 221, 500

*Includes counties and parishes listed in the preface.

Source: U.S. Census Bureau, *Fifteenth Census of the United States: 1930: Population*, Vol. 3: Part 1: 176–182, 974–979, 1274–1280, 1328–1337.

Table 5. Tilled Land and Farm Land in the Bottomlands (1880)

Regions*	Tilled Land (acres)**	Farm Land (acres)	Total Land (acres)
Mississippi Delta	428, 591	1, 191, 954	3, 865, 600
Louisiana Delta	368, 864	1, 042, 280	3, 046, 400
Arkansas Delta	608, 425	1, 850, 292	6, 438, 400
Missouri Delta	302, 666	692, 050	2, 124, 800
Totals	1, 708, 546	4, 776, 576	15, 475, 200

* Includes counties and parishes listed in the preface, except Humphreys County, Mississippi (created in 1918).

**Includes crop land as well as plowable pasture.

Sources: U.S. Census Office, *Tenth Census (1880): Report on the Productions of Agriculture* (Washington, D. C.: Government Printing Office, 1883), 139, 152–153, 156–158; U.S. Census Office, *Report on Cotton Production*, Part 1: 145–150, 318–323, 513–517, 585–596.

Table 6. Tilled Land and Farm Land in the Bottomlands (1930)

Regions*	Tilled Land (acres)**	Farm Land (acres)	Total Land (acres)
Mississippi Delta	2, 026, 483	2, 559, 620	3, 875, 840
Louisiana Delta	703, 613	1, 126, 280	3, 077, 120
Arkansas Delta	2, 406, 334	3, 194, 134	6, 211, 200
Missouri Delta	1, 247, 269	1, 448, 801	2, 102, 400
Totals	6, 383, 699	8, 328, 835	15, 266, 560

* Includes counties and parishes listed in the preface.
**Includes crop land as well as plowable pasture.

Source: U.S. Census Bureau, *Fifteenth Census for the United States: 1930: Agriculture*, Vol. 2: Part 1: 982–987; Part 2: 1050–1055, 1137–1141, 1215–1218.

Table 7. Cotton and Corn Production in the Bottomlands (1880)

Regions*	Cotton Land (acres)	Cotton (bales)	Corn Land (acres)	Corn (bushels)
Mississippi Delta	255, 638	197, 375	100, 613	2, 024, 845
Louisiana Delta	223, 348	184, 116	70, 720	1, 155, 199
Arkansas Delta	263, 569	183, 079	178, 453	3, 737, 161
Missouri Delta	23, 488	15,357	165, 086	5, 275, 619
Totals	766, 043	579, 927	514, 872	12, 192, 824

*Includes counties and parishes listed in the preface, except Humphreys County, Mississippi (created in 1918).

Source: U.S. Census Office, *Tenth Census (1880): Report on the Productions of Agriculture*, 213, 225–226, 229–231, 248, 261, 265–267.

Table 8. Cotton and Corn Production in the Bottomlands (1930)

Regions*	Cotton Land(acres)	Cotton (bales)	Corn Land (acres)	Corn (bushels)
Mississippi Delta	1, 534, 389	873, 659	222, 110	4, 533, 298
Louisiana Delta	405, 859	222, 274	129, 994	2, 424, 364
Arkansas Delta	1, 335, 041	662, 019	461, 144	9, 259, 852
Missouri Delta	330, 029	214, 037	382, 332	9, 965, 827
Totals	3, 605, 318	1, 971, 989	1, 195, 580	26, 183, 341

*Includes counties and parishes listed in the preface.

Sources: U.S. Census Office, *Fifteenth Census of the United States: 1930: Agriculture*, Vol. 2: Part 1: 1017–1023; Part 2: 1083–1089, 1169–1173, 1243–1247.

Table 9. Farm Tenancy and Ownership in the Bottomlands (1880)

Regions*	Tenant-Operated Farms**	Owner-Operated Farms	Total Farm Operators
Mississippi Delta	2, 043	2, 021	4, 064
Louisiana Delta	4, 460	1, 761	6, 221
Arkansas Delta	6, 763	6, 548	13, 311
Missouri Delta	3, 024	3, 944	6, 968
Totals	16, 290	14, 274	30, 564

* Includes counties and parishes listed in the preface, except Humphreys County, Mississippi (created in 1918).

**Includes cash renters, share renters, and sharecroppers.

Source: U. S. Census Office, *Tenth Census (1880): Report on the Productions of Agriculture*, 66–67, 92–93, 100–105.

Table 10. Farm Tenancy and Ownership in the Bottomlands (1930)

Regions*	Tenant-Operated Farms**	Owner-Operated Farms	Total Farm Operators
Mississippi Delta	82, 600	4, 770	87, 370
Louisiana Delta	25, 025	3, 480	28, 505
Arkansas Delta	61, 844	12, 665	74, 509
Missouri Delta	16, 317	4, 071	20, 388
Totals	185, 786	24, 986	210, 772

* Includes counties and parishes listed in the preface.
**Includes cash renters, share renters, and sharecroppers.

Source: U.S. Census Bureau, *Fifteenth Census of the United States: 1930: Agriculture*, Vol. 2: Part 1: 982–987; Part 2: 1050–1055, 1137–1141, 1215–1218.

Notes

PREFACE

1. Purificacion O. MacDonald and others, "Documentation, Chronology, and Future Projections of Bottomland Hardwood Habitat Loss in the Lower Mississippi Alluvial Plain" (Washington, D.C.: Division of Ecological Services, U.S. Department of the Interior, 1979), Vol. 1: 109.

CHAPTER 1

1. Donald H. Welsh, ed., "Martha Woods Visits Missouri in 1857," *Missouri Historical Review*, 55(January 1961): 111–112.

2. U.S. Census Office, *Tenth Census(1880): Report on Cotton Production in the United States* (Washington, D.C.: Government Printing Office, 1884), Part 1:87–88, 114–115, 241–246, 508, 549–554; Robert W. Harrison, *Alluvial Empire* (Little Rock, Ark.: Pioneer Press, 1961), Vol. 1:18–19, 43–48; Thomas Foti, "The River's Gifts and Curses," in Jeannie Whayne and Willard Gatewood, eds., *The Arkansas Delta: Land of Paradox* (Fayetteville, Ark.: University of Arkansas Press, 1993), 41–46.

3. "The Mississippi Swamp," *DeBow's Review*, o.s. 7(July 1849): 54–55; U.S. Census Office, *Report on Cotton Production*, Part 1: 85–86; Hugh H. Bennett, *The Soils and Agriculture of the Southern States* (New York, N.Y.: MacMillan Co., 1921), 315–319.

The loamy and clay soils of the bottomlands are now classified as Haplaquepts by soil scientists. See Soil Survey Staff, *Soil Taxonomy*, U.S. Department of Agriculture, Agriculture Handbook No. 436 (Washington, D.C.: Government Printing Office, 1975), 241–243.

4. "Mississippi," *DeBow's Review*, o.s. 7(July 1849): 42; U.S. Census Office,

Report on Cotton Production, Part 1: 86, 241, 245; William F. Gray, ed., *Imperial Bolivar* (Cleveland, Miss.: Bolivar Commercial, 1923), 7–8; E. Lucy Braun, *Deciduous Forests of Eastern North America* (New York, N.Y.: Free Press, 1974), 291–295; Harold K. Grimmett, "Early Plant and Animal Communities in the Arkansas Delta," *Arkansas Historical Quarterly,* 48(Summer 1989): 104–106.

5. Forest R. Moulton, ed., *A Symposium on Human Malaria,* American Association for the Advancement of Science, Publication No. 15 (Washington, D.C.: American Association for the Advancement of Science, 1941), 77, 84–85.

Antebellum settlers were understandably fearful of the malarial bottomlands. As an example, when Clark Ward's parents settled in eastern Arkansas Territory, they built "a good double log house" and "cleared 25 or 30 acres" of land, but they lost three children to "swamp fever," or malaria. Burying their children in the front yard, the Wards "decided to leave their home and move west" for "their health." Loading their possessions in an ox-drawn wagon, the Wards fled their malarial farmstead, leaving behind a "crib full of corn" as well as a "good field of green corn." See H. M. McIver, "Reminiscences of an Arkansas Pioneer as Recorded in 1890," *Arkansas Historical Quarterly,* 17(Spring 1958): 56.

6. See Willie D. Halsell, "Migration Into, and Settlement of, Leflore County, 1833–1876," *Journal of Mississippi History,* 9(October 1947): 221–223.

7. "Louisiana—Her Public Lands and Levees," *DeBow's Review,* o.s. 10(May 1851): 530; Robert Bowman, "Early History and Archaeology of Yazoo County," *Publications of the Mississippi Historical Society,* 8(Annual 1904): 433; Harrison, *Alluvial Empire,* Vol.1:67–70; Donald Holley, "The Plantation Heritage," in Whayne and Gatewood, eds., *The Arkansas Delta,* 239–240, 254.

8. Harrison, *Alluvial Empire,* Vol. 1:70–92; Leon P. Ogilvie, "Governmental Efforts at Reclamation in the Southeast Missouri Lowlands," *Missouri Historical Review,* 64(January 1970): 172.

9. Gray, ed., *Imperial Bolivar,* 8–10; Florence W. Sillers, comp., *History of Bolivar County, Mississippi* ed. by Wirt A. Williams (Jackson, Miss.: Hederman Bros., 1948), 9; John H. Moore, *The Emergence of the Cotton Kingdom in the Old Southwest,* 1770–1860 (Baton Rouge, La.: Louisiana State University Press, 1988), 128–129.

10. John F. Hart, "The Role of the Plantation in Southern Agriculture," in *Proceedings of the Tall Timbers Fire Ecology Conference No. 16* (Tallahassee, Fla.: Tall Timbers Research Station, 1982), 6–7; Moore, *Emergence of the Cotton Kingdom,* 116–117.

11. U.S. Census Office, *Population of the United States in 1860* (Washington, D.C.: Government Printing Office, 1864), 18, 194, 270, 286–287; Leon P. Ogilvie, "The Development of the Southeast Missouri Lowlands" (Ph.D. dissertation, University of Missouri, Columbia, 1967), 7, 30–31.

12. "The Parish of Tensas," *DeBow's Review,* o.s. 14(May 1853): 435; William A. Percy, *Lanterns on the Levee: Recollections of a Planter's Son* (New York, N.Y.: Alfred A. Knopf, 1941), 5; David A. Creed, "An Economic and Social History of Madison Parish, Louisiana" (M.A. thesis, Northeast Louisiana State College, Monroe, 1969), 12–13, 16–17; John C. Hudson, "The Yazoo–Mississippi Delta as Plantation Country," in *Proceedings of the Tall Timbers Fire Ecology Conference No. 16* (Tallahassee, Fla.: Tall Timbers Research Station, 1982), 67–69.

13. Frank A. Montgomery, *Reminiscences of a Mississippian in Peace and War* (Cincinnati, Ohio: Robert Clarke, 1901), 32.

14. U.S. Census Office, *Report on Cotton Production*, Part 1: 86–87; Gray, ed., *Imperial Bolivar*, 8; Harrison, *Alluvial Empire*, Vol. 1: 301.

15. Louis Hughes, *Thirty Years a Slave: From Bondage to Freedom* (Milwaukee, Wisc.: SouthSide Printing Co., 1897), 36, 77.

16. W. L. Murfree, "The Levees of the Mississippi," *Scribner's Monthly*, 22(July 1881): 423; Holley, "The Plantation Heritage," 241.

17. Montgomery, *Reminiscences of a Mississippian*, 32.

18. Frederick L. Olmsted, *A Journey in the BackCountry* (New York, N.Y.: Mason Bros., 1860), 72–82; Robert C. Reinders, "Dr. James G. Carson's Canebrake: A View of an Ante-Bellum Louisiana Plantation," *Louisiana Historical Quarterly*, 33(October 1950): 358; Merle Prunty, Jr., "The Renaissance of the Southern Plantation," *Geographical Review*, 45(October 1955): 465–466.

19. John Q. Anderson, ed., *Brokenburn: The Journal of Kate Stone, 1861–1868* (Baton Rouge, La.: Louisiana State University Press, 1955), xi, 3–6.

20. Walter Prichard, ed., "A Tourist's Description of Louisiana in 1860," *Louisiana Historical Quarterly*, 21(October 1938): 1208.

21. "The Parish of Concordia," *DeBow's Review*, o.s. 11(July 1851): 41, 59; T. Lynn Smith and S. Earl Grigsby, *The Situation and Prospects of the Population in the Black River Settlement, Louisiana*, Louisiana Agricultural Experiment Station, Bulletin No. 319 (Baton Rouge, La.: Louisiana State University, 1940), 4–6.

22. See Jack P. Oden, "A Century of Agriculture in the Carroll Parishes of Louisiana, 1800–1900," (M.A. thesis, Northeast Louisiana State College, Monroe, 1967), 26, 37–44; Ogilvie, "Development of the Southeast Missouri Lowlands," 26–31; Hart, "Role of the Plantation," 6–7; Jeannie M. Whayne, *A New Plantation South: Land, Labor and Federal Favor in Twentieth-Century Arkansas* (Charlottesville, Va.: University Press of Virginia, 1996), 10–12.

23. W. S. Kirby, *The Drummer Boy of the Ozarks; or, Sketches in the Life of Ben Elder* (West Plains, Mo.: Gazette Steam Printer, 1893), 5–6, 9–10.

24. "The Mississippi Swamp," 55; "Our Southern Tormentor—The Mosquito," *DeBow's Review*, o.s. 19(August 1855): 171–185; "Plantation Hygiene," *Southern Cultivator*, 15(May 1857): 140–141.

25. Dorothy Stanley, ed., *The Autobiography of Sir Henry Morton Stanley* (Boston, Mass.: Houghton Mifflin, 1937), 151, 155.

Quinine, a bitter alkaloid that was extracted from cinchona bark, eased malarial symptoms, but it cost about eight dollars an ounce before the Civil War. See Melinda S. Meade, "The Rise and Demise of Malaria," *Southeastern Geographer*, 20(November 1980): 86.

26. Col. Caleb G. Forshey, *The Delta of the Mississippi* (Cambridge, Mass.: John Wilson and Son, 1873), 83; Murfree, "The Levees of the Mississippi," 423–425.

27. "The Parish of Concordia," 60.

28. "The Delta of Northwestern Mississippi," *DeBow's Review*, o.s. 13(November 1852): 476; S. P. Reynolds, *History of Levees and Drainage in Southeastern Missouri*, Missouri Agricultural Experiment Station. Bulletin No. 323 (Columbia, Mo.: University of Missouri, 1933), 39; Harrison, *Alluvial Empire*, Vol.1:66–84; Arthur D. Frank, *The Development of the Federal Program of Flood Control on the Mississippi River* (New York, N.Y.: AMS Press, 1968), 26–27.

29. Walter Sillers, Sr., "Flood Control in Bolivar County, 1838–1924," *Journal of Mississippi History*, 9(January 1947): 5–6; Robert W. Harrison, "Levee Building in Mississippi Before the Civil War," *Journal of Mississippi History*, 12(April 1950): 80–82, 91–92.

30. See "The Parish of Tensas," 435; Halsell, "Leflore County, 1833–1876," 227–229; Oden, "Century of Agriculture in the Carroll Parishes," 46–47.

31. See Paul W. Gates, "The Railroads of Missouri, 1850–1870," *Missouri Historical Review*, 26(January 1932): 132, 138; Robert C. Black III, *The Railroads of the Confederacy* (Chapel Hill, N.C.: University of North Carolina Press, 1952), endpaper map; E. Dale Odom, "The Vicksburg, Shreveport and Texas: The Fortunes of a Scalawag Railroad," *Southwestern Social Science Quarterly*, 44(December 1963): 277; Leo E. Huff, "The Memphis and Little Rock Railroad During the Civil War," *Arkansas Historical Quarterly*, 23(Autumn 1964): 260.

32. R. D. Calhoun, "A History of Concordia Parish, Louisiana," II, *Louisiana Historical Quarterly*, 15(April 1932): 224; Sillers, comp., *History of Bolivar County*, 7–9; Mabel F. Edrington, *History of Mississippi County, Arkansas* (Ocala, Fla.: Ocala Star-Banner, 1962), 30; Betty F. Powell, *History of Mississippi County, Missouri* (Independence, Mo.: BNL Library Service, 1975), 52–53; Carl H. Moneyhon, "Economic Democracy in Antebellum Arkansas, Phillips County, 1850–1860," *Arkansas Historical Quarterly*, 40(Summer 1981): 159–160.

33. The impact of the Civil War (1861–1865) on the Southern bottomlands is discussed in Marshall Dial, *The Bootheel Swamp Struggle* (Lilbourn, Mo.: Lloyd Publications, 1961); John D. Winters, "Secession and Civil War in Northern Louisiana," in B. H. Gilley, ed., *North Louisiana* (Ruston, La.: McGinty Trust Fund Publications, 1984), 159–193; James C. Cobb, *The Most Southern Place on Earth: The Mississippi Delta and the Roots of Regional Identity* (New York, N.Y.: Oxford University Press, 1992), 29–46; Bobby Roberts, "Desolation Itself: The Impact of the Civil War," in Whayne and Gatewood, eds., *The Arkansas Delta*, 70–93.

34. Anderson, ed., *Brokenburn*, 191.

35. See Thomas W. Knox, *Camp-Fire and Cotton-Field: Southern Adventures in Time of War* (New York, N.Y.: Blelock and Co., 1865), 311, 342; Lawrence N. Powell, *New Masters: Northern Planters During the Civil War and Reconstruction* (New Haven, Conn.: Yale University Press, 1980), 6–7, 9; James T. Currie, *Enclave: Vicksburg and Her Plantations, 1863–1870* (Jackson, Miss.: University Press of Mississippi, 1980), 56–61; Ronald L. F. Davis, *Good and Faithful Labor: From Slavery to Sharecropping in the Natchez District, 1860–1890* (Westport, Conn.: Greenwood Press, 1982), 60–73.

36. Knox, *Camp-Fire and Cotton-Field*, 313, 443–449; Anderson, ed., *Brokenburn*, 298; D. Clayton James, "Mississippi Agriculture, 1861–1865," *Journal of Mississippi History*, 24(July 1962): 139; Currie, *Enclave*, 62–67; Powell, *New Masters*, 46; Roberts, "Desolation Itself," 86.

37. Gray, ed., *Imperial Bolivar*, 14; Harrison, *Alluvial Empire*, Vol. 1:96–97; James, "Mississippi Agriculture, 1861–1865," 135–136; Joel G. Taylor, *Louisiana Reconstructed, 1863–1877* (Baton Rouge, La.: Louisiana State University Press, 1974), 315, 319; Carl H. Moneyhon, "The Impact of the Civil War in Arkansas: The Mississippi River Plantation Counties," *Arkansas Historical Quarterly*, 51(Summer 1992): 110–111.

38. W. S. Odlham, "Last Days of the Confederacy," *DeBow's Review*, a.w.s. 8(August 1870): 651.

39. Montgomery, *Reminiscences of a Mississippian*, 258–260.

40. Louis S. Gerteis, *From Contraband to Freedman* (Westport, Conn.: Greenwood Press, 1973), 187; Michael Wayne, *The Reshaping of Plantation Society: The Natchez District, 1860–1880* (Baton Rouge, La.: Louisiana State University Press, 1983), 35–38; Carl H. Moneyhon, *The Impact of the Civil War and Reconstruction in Arkansas* (Baton Rouge, La.: Louisiana State University Press, 1994), 212–213.

41. "A Victim" [George C. Benham], *A Year of Wreck: A True Story* (New York, N.Y.: Harper and Bros., 1880), 11, 15, 71; Powell, *New Masters*, 38, 41.

42. "Southern Estates in the Market," *DeBow's Review*, a.w.s. 2(December 1866): 667.

43. Whitelaw Reid, *After the War: A Southern Tour* (Cincinnati, Ohio: Moore, Wilstach and Baldwin, 1866), 455, 480–481; A. T. Morgan, *Yazoo; or, on the Picket Line of Freedom in the South* (Washington, D.C.: Press of Rufus Darby, 1884), 15, 18–20, 26, 29; Powell, *New Masters*, 42–43.

44. Robert M. Davis, *The Southern Planter, the Factor, and the Banker* (New Orleans, La.: Robert M. Davis, 1871), 6, 12–14; Anderson, ed., *Brokenburn*, 368; Powell, *New Masters*, 39–40.

45. B. F. Nourse, "Report on Cotton," *DeBow's Review*, a.w.s. 7(September 1869): 783–791; Anderson, ed., *Brokenburn*, 369; Taylor, *Louisiana Reconstructed*, 401; Wayne, *Reshaping of Plantation Society*, 63–68.

46. Charles H. Brough, "History of Taxation in Mississippi," *Publications of the Mississippi Historical Society*, 3(Annual 1899): 121–123; Roger W. Shugg, "Survival of the Plantation System in Louisiana," *Journal of Southern History*, 3(August 1937): 317; William I. Hair, *Bourbonism and Agrarian Protest: Louisiana Politics, 1877–1900* (Baton Rouge, La.: Louisiana State University Press, 1969), 100; Wayne, *Reshaping of Plantation Society*, 78, 84.

47. William E. Highsmith, "Louisiana Landholding During War and Reconstruction," *Louisiana Historical Quarterly*, 38(January 1955): 53–54; Taylor, *Louisiana Reconstructed*, 358–360; Davis, *Good and Faithful Labor*, 126; Wayne, Reshaping of Plantation Society, 80–84; Willard B. Gatewood, Jr., "Sunnyside: The Evolution of an Arkansas Plantation, 1840–1945," *Arkansas Historical Quarterly*, 50(Spring 1991):12–15.

48. Robert Somers, *The Southern States Since the War, 1870–1* (London, U.K.: MacMillan and Co., 1871), 256–257; Halsell, "Leflore County, 1833–1876," 235–236; Davis, *Good and Faithful Labor*, 129, 143, 145; Wayne, *Reshaping of Plantation Society*, 91–99; Moneyhon, "Impact of the Civil War in Arkansas," 115.

49. Edward King, *The Great South: A Record of Journeys* (Hartford, Conn.: American Publishing Co., 1875), 273; Charles Nordhoff, *The Cotton States in the Spring and Summer of 1875* (New York, N.Y.: D. Appleton and Co., 1876), 38.

50. A. B. Hurt, *Mississippi*, U.S. Department of Agriculture, Miscellaneous Special Report No. 3 (Washington, D.C.: Government Printing Office, 1883), 30–31.

51. King, *The Great South*, 298–300; Nordhoff, *Cotton States*, 21, 37–38, 72, 84; Willie D. Halsell, "Migration Into, and Settlement of, Leflore County in the Later Periods, 1876–1920," *Journal of Mississippi History*, 10(July 1948): 243–244; Taylor, *Louisiana Reconstructed*, 327, 329, 372–376; Fon Louise Gordon, "From Slav-

ery to Uncertain Freedom: Blacks in the Delta," in Whayne and Gatewood, eds., *The Arkansas Delta*, 98–99, 104.

52. John C. Willis, "On the New South Frontier: Life in the Yazoo–Mississippi Delta, 1865–1920" (Ph.D. dissertation, University of Virginia, Charlottesville, 1991), 67–80.

53. "Plantations in the South," in U.S. Census Bureau, *Thirteenth Census of the United States Taken in the Year 1910: Agriculture* (Washington, D.C.: Government Printing Office, 1913), Vol. 5:877–878.

54. See Albert B. Hart, *The Southern South* (New York, N.Y.: D. Appleton and Co., 1910), 255–256; Prunty, "Renaissance of the Southern Plantation," 466–482.

55. Lewis F. Thomas, "Basic Factors in Flood Frequency in the Lower Mississippi River," *Annals of the American Academy of Political and Social Science*, 135(January 1928): 1.

56. "The Louisiana Levees," *DeBow's Review*, a.w.s. 3(April–May 1867): 469, 471; "The Levees of the Mississippi," *DeBow's Review*, a.w.s. 5(February 1868): 163; *Letter of Gov. Jas. L. Alcorn and Resolution Memorializing Congress Relative to Overflowed Lands on the Mississippi River* (Jackson, Miss.: Kimball, Raymond and Co., 1870), 1, 5–7.

57. Reynolds, *History of Levees and Drainage in Southeast Missouri*, 39.

58. Thomas S. Staples, *Reconstruction in Arkansas, 1862–1874*, Columbia University, Studies in History, Economics and Public Law, No. 245 (New York, N.Y.: Columbia University, 1923), 344–347; Robert W. Harrison and Walter M. Kollmorgen, "Land Reclamation in Arkansas Under the Swamp Land Grant of 1850," *Arkansas Historical Quarterly*, 61(Winter 1947): 408–409.

59. Sam Mims, "Louisiana's Administration of Swamp Land Funds," *Louisiana Historical Quarterly*, 28(January 1945): 305–310; Taylor, *Louisiana Reconstructed*, 86, 140, 193–196.

60. Robert W. Harrison, *Levee Districts and Levee Building in Mississippi* (Stoneville, Miss.: Delta Council and others, 1951), 27–29, 35–37; Harrison, *Alluvial Empire*, Vol.1:97–101.

61. Arthell Kelley, "Levee Building and the Settlement of the Yazoo Basin," *Southern Quarterly*, 1(July 1963): 298; Lillian Pereyra, "James Lusk Alcorn and a Unified Levee System," *Journal of Mississippi History*, 27(February 1965): 38.

62. Sillers, "Flood Control in Bolivar County," 9–10; Harrison, *Alluvial Empire*, Vol.1:102–103; Frank, *Federal Program of Flood Control*, 41–42, 116; Mary G. McBride and Ann M. McLaurin, "The Origin of the Mississippi River Commission," *Louisiana History*, 36(Fall 1995): 401–402.

63. "Down the Mississippi," *Harper's New Monthly Magazine*, 41(November 1870): 835–844; Louis C. Hunter, *Steamboats on the Western Rivers* (New York, N.Y.: Octagon Books, 1969), 220–222, 585, 612–613, 633.

64. Edward H. Hall, *Appleton's Hand-Book of American Travel: The Southern Tour* (New York, N.Y.: D. Appleton and Co., 1872), 386–387, 400, 407–408; Frederick W. Williamson and George T. Goodman, eds., *Eastern Louisiana* (Monroe, La.: Historical Record Association, 1939), Vol.1:220–221; Halsell, "Lefore County in Later Periods," 241–242; John D. Winters, "The Ouachita-Black," in Edwin A. Davis, ed., *The Rivers and Bayous of Louisiana* (Baton Rouge, La.: Louisiana Educational Research Association, 1968), 27.

65. U.S. Congress, House of Representatives, *St. Francis River, Mo.[Missouri]*

and Ark.[Arkansas], 71st Congress, 2nd Session, House Document No. 159 (Washington, D.C.: Government Printing Office, 1930), 17; Robert L. Brandfon, *Cotton Kingdom of the New South: A History of the Yazoo–Mississippi Delta from the Reconstruction to the Twentieth Century* (Cambridge, Mass.: Harvard University Press, 1967), 66–67; Nollie W. Hickman, "The Tensas–Bayou Macon," in Davis, ed., *The Rivers and Bayous of Louisiana*, 19–20; Harry P. Owens, *Steamboats and the Cotton Economy: River Trade in the Yazoo–Mississippi Delta* (Jackson, Miss.: University Press of Mississippi, 1990), 17, 72–73.

66. Huff, "The Memphis and Little Rock Railroad," 260–265; Odom, "The Vicksburg, Shreveport and Texas," 277; Ogilvie, "Development of the Southeast Missouri Lowlands," 37–38.

67. Williamson and Goodman, eds., *Eastern Louisiana*, Vol.1:249–251; Ira G. Clark, *Then Came the Railroads: The Century from Steam to Diesel in the Southwest* (Norman, Okl.: University of Oklahoma Press, 1958), 51; Odom, "The Vicksburg, Shreveport and Texas," 278–285; Taylor, *Louisiana Reconstructed*, 340.

68. James P. Henry, *The Resources of the State of Arkansas* (Little Rock, Ark.: Price and McClure, 1873), 68, 80–81; Clark, *Then Came the Railroads*, 48–49; Carl H. Moneyhon, "Delta Towns," in Whayne and Gatewoods, eds., *The Arkansas Delta*, 212–213.

69. Lee A. Dew and Louis Koeppe, "Narrow-Gauge Railroads in Arkansas," *Arkansas Historical Quarterly*, 31(autumn 1972): 277; James W. Leslie, *Land of the Cypress and Pine* (Little Rock, Ark.: Rose Publishing Co., 1976), 67–83.

70. Robert S. Douglass, *History of Southeast Missouri* (New York, N.Y.: Lewis Publishing Co., 1912), Vol. 1:499–500; Ogilvie, "Development of the Southeast Missouri Lowlands," 40–43.

71. Brandfon, *Cotton Kingdom of the New South*, 68; Willis, "On the New South Frontier," 206–222; Cobb, *Most Southern Place on Earth*, 80.

72. See John S. Otto, *The Southern Frontiers*, 1607–1860 (Westport, Conn.: Greenwood Press, 1989) 2–3.

CHAPTER 2

1. Wm. H. Harris, *Louisiana Products, Resources and Attractions* (New Orleans, La.: New Orleans Democrat Printer, 1881), 135.

2. See U.S. Census Office, *Tenth Census (1880): Report on the Productions of Agriculture* (Washington, D.C.: Government Printing Office, 1883); U.S. Census Office, *Tenth Census (1880): Report on Cotton Production in the United States* (Washington, D.C.: Government Printing Office, 1884), Parts 1–2.

3. U.S. Census Office, *Report on the Production of Agriculture*, 41, 265; U.S. Census Office, *Report on Cotton Production*, Part 1: 19, 321, 323.

4. U.S. Census Office, *Report on the Production of Agriculture*, 41, 261; U.S. Census Office, *Report on Cotton Production*, Part 1:20, 145, 149.

5. U.S. Census Office, *Report on the Production of Agriculture*, 41, 248; U.S. Census Office, *Report on Cotton Production*, Part 1:20, 585.

6. U.S. Census Office, *Report on the Production of Agriculture*, 41, 266–267; U.S. Census Office, *Report on Cotton Production*, Part 1:21, 508–509, 513–517; Oliver E. Baker, "Agricultural Regions of North America: Part I: The Basis of Classification,"*Economic Geography*, 2(July 1926): 473.

7. U.S. Census Office, *Report on the Production of Agriculture*, 37, 139, 152–153, 156–158.

8. See *The West-Side Delta* (Memphis, Tenn.: Southern Alluvial Land Association, 1920), 23; Willie D. Halsell, "Migration Into, and Settlement of, Leflore County, 1833–1876,"*Journal of Mississippi History*, 9(October 1947): 231–232; Leon P. Ogilvie, "The Development of the Southeast Missouri Lowlands" (Ph.D. dissertation, University of Missouri, Columbia, 1967), 177–178.

9. U.S. Census Office, *Report on Cotton Production*, Part 1:241, 243–244, 275

10. U.S. Census Office, *Report on Cotton Production*, Part 1:145–150, 275, 318–323, 513–517, 585–594.

11. U.S. Congress, Senate, Committee on Education and Labor, *Report of the Committee of the Senate Upon the Relations Between Labor and Capital, and Testimony Taken by the Committee* (Washington, D.C.: Government Printing Office, 1885), Vol. 2:158, 168, 170–171; Willie D. Halsell, "Migration Into, and Settlement of, Leflore County in the Later Periods, 1876–1920," *Journal of Mississippi History*, 10(July 1948): 243; Fon Louise Gordon, "From Slavery to Uncertain Freedom," in Jeannie Whayne and Willard Gatewood, eds., *The Arkansas Delta* (Fayetteville, Ark.: University of Arkansas Press, 1993), 99.

12. U.S. Census Office, *Population of the United States in 1860* (Washington, D.C.: Government Printing Office, 1864), 18, 194, 270, 286–287; U.S. Census Office, *Statistics of the Population of the United States at the Tenth Census* (Washington, D.C.: Government Printing Office, 1883), 381, 393–394, 397–399.

13. See John C. Willis, "On the New South Frontier: Life in the Yazoo–Mississippi Delta, 1865–1920" (Ph.D. dissertation, University of Virginia, Charlottesville, 1991), 106–120.

14. U.S. Census Office, *Report on Cotton Production*, Part 1:146, 513–514, 586.

15. U.S. Congress, *Report of the Committee of the Senate*, Vol. 2:169; Dunbar Rowland, *History of Mississippi* (Chicago, Ill.: S. J. Clarke, 1925), Vol. 2:208–213; O. V. Wells, "The Depression of 1873–79," *Agricultural History*, 11(July 1937): 239–247.

16. See *"Boost Missouri" and "The Call of the Farm"* (Jefferson City, Mo.: Missouri Bureau of Labor Statistics, [1915]), 11; William F. Gray, ed., *Imperial Bolivar* (Cleveland, Miss.: Bolivar Commercial, 1923), 18.

An Arkansan, who managed his uncle's indebted plantation in Monroe County, paid off much of the debt by selling gum trees to a railway company. Paying fifty cents per tree, the company transformed gum logs into durable railroad ties. See Sue Ellen P. McDowall, *Cotton and Jasmine: A Southern Mosaic* (New York, N.Y.: Vantage Press, 1956), 23.

17. E. G. Wall, *Hand-Book of the State of Mississippi* (Jackson, Miss.: Mississippi State Board of Immigration and Agriculture, 1885), 12–13; *Arkansas: Statistics and Information* (Little Rock, Ark.: Missouri Pacific Railway Co., 1888), 17–18; Wm. C. Stubbs, *A Hand-Book of Louisiana* (New Orleans, La.: New Orleans Picayune, 1895), 52; Leon P. Ogilvie, "Governmental Efforts at Reclamation in the Southeast Missouri Lowlands," *Missouri Historical Review*, 64(January 1970): 169.

18. James P. Henry, *Resources of the State of Arkansas* (Little Rock, Ark.: Price and McClure, 1873), 80–81; *Arkansas: Statistics and Information*, 19; Ira G. Clark, *Then Came the Railroads* (Norman, Okl.: University of Oklahoma Press, 1958), 34–

35; E. Dale Odom, "The Vicksburg, Shreveport and Texas," *Southwestern Social Science Quarterly*, 44(December 1963): 278, 283.

19. Robert L. Brandfon, *Cotton Kingdom of the New South* (Cambridge, Mass.: Harvard University Press, 1967), 46–67.

20. Brandfon, *Cotton Kingdom of the New South*, 70–72, 81–82; John C. Hudson, "The Yazoo–Mississippi as Plantation Country," in *Proceedings of the Tall Timbers Fire Ecology Conference No. 16* (Tallahassee, Fla.: Tall Timbers Research Station, 1982), 73–74, 76–79; Willis, "On the New South Frontier," 221–224.

21. See Frederick W. Williamson and George T. Goodman, eds., *Eastern Louisiana* (Monroe, La.: Historical Record Association, 1939), Vol. 1:251–254; Clark, *Then Came the Railroads*, 133, 138–140, 157–158; Clifton Hull, *Shortline Railroads of Arkansas* (Norman, Okl.: University of Oklahoma Press, 1969), 384; Lee A. Dew and Louis Koeppe, "Narrow-Gauge Railroads in Arkansas," *Arkansas Historical Quarterly*, 31(Autumn 1972): 279–291; James W. Leslie, *Land of the Cypress and Pine* (Little Rock, Ark.: Rose Publishing Co., 1976), 87–97; Leon P. Ogilvie, "The Development of the Southeast Missouri Lowlands" (Ph.D. dissertation, University of Missouri, Columbia, 1967), 43.

22. Hon. Isaiah T. Montgomery, "In the Yazoo–Mississippi Delta,"*Southern Workman*, 29(October 1900): 645.

23. Octave Thanet, "Town Life in Arkansas," *Atlantic Monthly*, 68(September 1891): 333.

24. Gray, ed., *Imperial Bolivar*, 41; Florence W. Sillers, comp., *History of Bolivar County, Mississippi* ed. by Wirt A. Williams (Jackson, Miss.: Hederman Bros., 1948), 282–283; 443; Robert Palmer, *Deep Blues* (New York, N.Y.: Viking Press, 1981), 52–54.

25. Halsell, "Leflore County in the Later Periods," 242–243; Irvin G. Wyllie, "Race and Class Conflict on Missouri's Cotton Frontier," *Journal of Southern History*, 20(May 1954): 184; Arthell Kelley, "Sequent Occupancy of the Yazoo Basin, Mississippi, 1830 to 1976," *Mississippi Geographer*, 6(Spring 1978): 8–9; Jeannie Whayne, "Creation of a Plantation System in the Arkansas Delta in the Twentieth Century," *Agricultural History*, 66(Winter 1992): 63–64, 66–68.

26. Gray, ed., *Imperial Bolivar*, 18; Halsell, "Leflore County in the Later Periods," 249–250; Nollie W. Hickman, "Mississippi Forests," in Richard A. McLemore, ed., *A History of Mississippi* (Jackson, Miss.: University Press of Mississippi, 1973), Vol.2:220–221; Bonnie Stepenoff, "The Last Tree Cut Down: The End of the Bootheel Frontier, 1880–1940,"*Missouri Historical Review*, 90(October 1995): 64–65.

27. Sillers, comp., *History of Bolivar County*, 170, 174, 407.

28. William L. Giles, "Agricultural Revolution, 1890–1970," in McLemore, ed., *A History of Mississippi*, Vol. 2:184.

29. Sillers, comp., *History of Bolivar County*, 170, 175, 407.

30. See Halsell, "Leflore County in the Later Periods," 249–250; Donald Holley, "The Plantation Heritage," in Whayne and Gatewoods, eds., *The Arkansas Delta*, 253; Jeannie Whayne, *A New Plantation South* (Charlottesville, Va.: University Press of Virginia, 1996), 24–27.

31. Sydney Nathans, "Gotta Mind to Move, A Mind to Settle Down: Afro-Americans and the Plantation Frontier," in William J. Cooper, Jr., and others, eds.,

A Master's Due: Essays in Honor of David Herbert Donald (Baton Rouge, La.: Louisiana State University Press, 1985), 213.

32. William Pickens, *Bursting Bonds: The Autobiography of a "New Negro"* ed. by William L. Andrews (Bloomington, Ind.: Indiana University Press, 1991), 4–11.

33. U.S. Congress, *Report of the Committee of the Senate*, Vol. 2:170.

34. See James C. Cobb, *The Most Southern Place on Earth* (New York, N.Y.: Oxford University Press, 1992), 82–83; Gordon, "From Slavery to Uncertain Freedom," and Holley, "The Plantation Heritage," in Whayne and Gatewood, eds., *The Arkansas Delta*, 98, 250–251.

35. U.S. Congress, *Report of the Committee of the Senate*, Vol. 2:158–160.

36. Nathans, "Gotta Mind to Move," 209–210. See also Willis, "On the New South Frontier," 120.

37. Montgomery, "In the Yazoo–Mississippi Delta," 645–647.

38. U.S. Congress, *Report of the Committee of the Senate*, Vol. 2:163, 168.

39. Ray S. Baker, *Following the Color Line: An Account of Negro Citizenship in the American Democracy* (New York, N.Y.: Doubleday, Page and Co., 1908), 102–103.

40. Octave Thanet, "Plantation Life in Arkansas," *Atlantic Monthly*, 68(July 1891): 35.

41. McDowall, *Cotton and Jasmine*, 29.

42. Lewis, F. Thomas, "Basic Factors in Flood Frequency in the Lower Mississippi River," *Annals of the American Academy of Political and Social Sciences*, 135(January 1928): 1.

43. Frank H. Tompkins, *Riparian Lands of the Mississippi River* (New Orleans, La.: F. H. Tompkins, 1901), 295–306; John L. Mathews, *Remaking the Mississippi* (Boston, Mass.: Houghton, Mifflin Co., 1909), 64–65; Robert W. Harrison, *Levee Districts and Levee Building in Mississippi* (Stoneville, Miss.: Delta Council and others, 1951), 51–52, 61–90; Robert W. Harrison, "The Formative Years of the Yazoo–Mississippi Delta Levee District," *Journal of Mississippi History*, 13(October 1951): 236–247.

44. Tompkins, *Riparian Lands of the Mississippi River*, 309–344; Sam Mims, "Louisiana's Administration of Swamp Land Funds," *Louisiana Historical Quarterly*, 28(January 1945): 312–313; Robert W. Harrison and Walter M. Kollmorgen, "Land Reclamation in Arkansas Under the Swamp Land Grant of 1850," *Arkansas Historical Quarterly*, 6(Winter 1947): 413–414.

45. Tompkins, *Riparian Lands of the Mississippi River*, 22, 53–55, 78, 80, 96, 109, 130, 133–135; Mathews, *Remaking the Mississippi*, 60–66.

46. J. N. Darling, "The Yazoo Delta of Mississippi and Location and Construction of Its Railroads," *Journal of the Western Society of Engineers*, 7(December 1902): 563–569.

47. S. M. Tracy, *Mississippi As It Is: A Handbook of Facts for Immigrants* (Jackson, Miss.: Messenger Publishing Co., 1895), 23; Brandfon, *Cotton Kingdom of the New South* 79–83; John F. Stover, *History of the Illinois Central Railroad* (New York, N.Y.: MacMillan Co., 1975), 169–171, 226–229; Burke Davis, *The Southern Railway: Road of Innovators* (Chapel Hill, N.C.: University of North Carolina Press, 1985), 191–192; Cobb, *Most Southern Place on Earth*, 81.

48. *In Invitation to Immigrants: Louisiana* (Baton Rouge, La.: Louisiana Bureau of Immigration, 1894), 81; Tompkins, *Riparian Lands of the Mississippi*, 407; R. D.

Calhoun, "History of Concordia Parish, Louisiana," VII, *Louisiana Historical Quarterly*, 16(July 1933): 471–472.

49. *Arkansas: Statistics and Information*, 6–7; Tompkins, *Riparian of the Mississippi*, 407; Stephen E. Wood, "The Development of Arkansas Railroads: Part II," *Arkansas Historical Quarterly*, 7(Autumn 1948): 164–165, 172; Clark, *Then Came the Railroads*, 238–239; Hull, *Shortline Railroads of Arkansas*, 384.

50. Tompkins, *Riparian Lands of the Mississippi*, 385, 388; William T. Doherty, Jr., *Louis Houck, Missouri Historian and Entrepreneur*, University of Missouri Studies, Vol. 33 (Columbia, Mo.: University of Missouri Press, 1960), 38–51; Ogilvie, "Development of the Southeast Missouri Lowlands," 44–56; Stepenoff, "Last Tree Cut Down," 64–65.

51. Charles Hoffmann, "The Depression of the Nineties," *Journal of Economic History*, 16(June 1956): 152–156; Willis, "On the New South Frontier," 260–261.

52. Industrial Commission, *Report of the Industrial Commission on Agriculture and Agricultural Labor, Including Testimony, with Review and Topical Digest Thereof* (Washington, D.C.: Government Printing Office, 1900–1902), Vol. 19:63, 140; U.S. Census Office, *Twelfth Census of the United States, Taken in the Year 1900: Agriculture* (Washington, D.C.: Government Printing Office, 1902), Part 2:407; George K. Holmes. *Cotton Crop of the United States*, 1790–1911. U.S. Department of Agriculture, Bureau of Statistics, Circular No. 32. (Washington, D.C.: Government Printing Office, 1912), 8.

53. Industrial Commission, *Report*, Vol. 10:464–466; Vol. 19:140–141.

54. Industrial Commission, *Report*, Vol. 10: 465, 485–486. See also, Alfred H. Stone, "The Negro Farmer in the Mississippi Delta," *Southern Workman*, 32(October 1903): 460.

55. Alfred H. Stone, *Studies in the American Race Problem* (New York, N.Y.: Doubleday, Page and Co., 1908), 113.

56. Willis, "On the New South Frontier," 9.

57. Alfred H. Stone, "The Negro in the Yazoo–Mississippi Delta," *Publications of the American Economic Association*, 3, No. 1(1902):255.

58. Stone, "The Negro Farmer in the Mississippi Delta," 458; Stone, *American Race Problem*, 128–129.

59. Nathans, "Gotta Mind to Move," 214.

60. See *Arkansas* (St. Louis, Mo.: Woodward and Tiernan, 1890); *Missouri: Imperial Mistress of States* (Columbia, Mo.: Missouri State Board of Agriculture, 1898); *An Invitation to Immigrants: Louisiana*; Tracy, *Mississippi As It Is.*

61. "Immigration, *Southern Farm Magazine*, 6(June 1898): 17; Industrial Commission, *Report*, Vol. 10:486; Hudson, "The Yazoo–Mississippi Delta," 79–82.

62. Industrial Commission, *Report*, Vol. 10:493, 496; Byrd Gibbens, "Strangers in the Arkansas Delta: Ethnic Groups and Nationalities," in Whayne and Gatewood, eds., *The Arkansas Delta*, 163–166.

63. Willard B. Gatewood, Jr., "Sunnyside: The Evolution of an Arkansas Plantation, 1840–1945," *Arkansas Historical Quarterly*, 50(Spring 1991): 23–26.

64. Willis, "On the New South Frontier," 50–52; Gibbens, "Strangers in the Arkansas Delta," 173–174.

65. Tompkins, *Riparian Lands of the Mississippi*, 385, 422, 499.

66. Halsell, "Leflore County in the Later Periods," 249–251; Wyllie, "Race and

Class Conflict on Missouri's Cotton Frontier," 184–185; Stepenoff, "Last Tree Cut Down," 66.

67. Nannie S. Jackson, *Vinegar Pie and Chicken Bread: A Woman's Diary of Life in the Rural South, 1890–1891* ed. by Margaret J. Bolsterli (Fayetteville, Ark.: University of Arkansas Press, 1982), 8–19, 50, 91–92.

68. "Letters from Prominent Cotton-Growers," in U.S. Congress, Senate, Committee on Agriculture and Forestry, *Report of the Committee on Agriculture and Forestry on Condition of Cotton Growers in the United States*, 53rd U.S. Congress, 3rd Session, Senate Report No. 986 (Washington, D.C.: Government Printing Office, 1907), Part 1:334.

69. U.S. Census Office, *Twelfth Census of the United States, Taken in the Year 1900: Population* (Washington, D.C.: Government Printing Office, 1901), Part 1: 10–11, 22–23, 26–28; Tompkins, *Riparian Lands of the Mississippi*, 351–352.

70. John S. Otto, *Southern Frontiers*, 1607–1860 (Westport, Conn.: Greenwood Press, 1989), 3.

71. John S. Otto, *Southern Agriculture During the Civil War-Era, 1860–1880* (Westport, Conn.: Greenwood Press, 1994), 79–81.

72. J. Crawford King, Jr., "The Closing of the Southern Range: An Exploratory Study," *Journal of Southern History*, 48(February 1982): 59, 64, 69.

73. Earl W. Hayter, "Barbed Wire Fencing—A Prairie Invention," *Agricultural History*, 13(October 1939):189–190, 195; Martin L. Primack, "Farm Fencing in the Nineteenth Century," *Journal of Economic History*, 29(June 1969): 288–291; King, "Closing of the Southern Range," 57.

74. King, "Closing of the Southern Range," 60.

Similar "stock laws" were also adopted in Arkansas, Louisiana, and Missouri. An Arkansas statute of 1891 permitted county courts to create fencing districts within counties. In turn, a Louisiana statute of 1870 permitted the police juries of parishes to enact "no-fence" laws for wards within parishes. Finally, a Missouri statute of 1883 required stock owners to enclose animals on their property, if the majority of voters in a county agreed to close the range. See *Arkansas Digest: 1820 to Date* (St. Paul, Minn.: West Publishing Co., 1970), Vol. 9:227–229; *West's Louisiana Statutes Annotated* (St. Paul, Minn.: West Publishing Co., 1988), Vol. 19:495; *Missouri Digest: 1821 to Date* (St. Paul Minn.: West Publishing Co., 1952), Vol. 2: 484–485.

75. See D. E. Salmon, *Hog Cholera and Swine Plague*, U.S. Department of Agriculture, Farmers' Bulletin No. 24 (Washington, D.C.: Government Printing Office, 1894), 3–5; John R. Mohler, *Texas or Tick Fever and Its Prevention*, U.S. Department of Agriculture, Farmers' Bulletin No. 258 (Washington, D.C.: Government Printing Office, 1906), 5–8; Julius Rubin, "The Limits of Agricultural Progress in the Nineteenth-Century South," *Agricultural History*, 49(April 1975): 366–367; Tamara M. Haygood, "Cows, Ticks and Disease: A Medical Interpretation of the Southern Cattle Industry,"*Journal of Southern History*, 52(November 1986): 562–563.

CHAPTER 3

1. *About the South on Lines of the Illinois Central and Yazoo & Mississippi Valley Railroad* (Chicago, Ill.: Illinois Central Railroad Co., 1904), 37.

2. Frank H. Tompkins, *Riparian Lands of the Mississippi River* (New Orleans, La.: F. H. Tompkins, 1901), 295–344; Robert W. Harrison, *Levee Districts and Levee Building in Mississippi* (Stoneville, Miss.: Delta Council and others, 1951), 134, 140, 148–149, 154, 163–169, 172, 182–183, 188, 192, 199–200.

3. *About the South*, 8,11; J. O. Wright, *Swamp and Overflowed Lands in the United States: Ownership and Reclamation*, U.S. Department of Agriculture, Office of Experiment Stations, Circular No. 76 (Washington, D.C.: Government Printing Office, 1907), 10–11.

4. H. S. Yohe, *Organization, Financing and Administration of Drainage Districts*, U.S. Department of Agriculture, Farmers' Bulletin No. 815 (Washington, D.C.: Government Printing Office, 1917), 7, 9–11, 13; S. H. McCrory, "The Drainage Movement in the United States," in U.S. Department of Agriculture, *Yearbook of the U.S. Department of Agriculture: 1918* (Washington, D.C.: Government Printing Office, 1919), 138–140.

5. J. O. Wright, *Excavating Machinery Used for Digging Ditches and Building Levees*, U.S. Department of Agriculture, Office of Experiment Stations, Circular No. 74 (Washington, D.C.: Government Printing Office, 1907), 5–7, 29–35; A. E. Morgan and others, *A Preliminary Report on the Drainage of the Fifth Louisiana Levee District*, U.S. Department of Agriculture, Office of Experiment Station, Circular No. 104 (Washington, D.C.: Government Printing Office, 1911), 27; J. Sheppard Smith, "Reclamation of Swamp Lands and the Modern Drainage Bond," *Annals of the American Academy of Political and Social Science*, 88(March 1920): 103–109.

6. Robert W. Harrison, *Alluvial Empire* (Little Rock, Ark.: Pioneer Press, 1961), Vol. 1:217; Robert L. Brandfon, *Cotton Kingdom of the New South* (Cambridge, Mass.: Harvard University Press, 1967), 125.

7. Roger D. Marsden and R. P. Teele, *Economic Status of Drainage Districts in the South in 1926*, U.S. Department of Agriculture, Technical Bulletin No. 194 (Washington, D.C.: Government Printing Office, 1930), 4; S. E. Simonson, "Origin of Drainage Projects in Mississippi County," *Arkansas Historical Quarterly*, 5(Fall 1946): 264–272.

8. Marsden and Teele, *Economic Status of Drainage Districts*, 4,10; Smith, "Reclamation of Swamp Lands," 110–111; Bonnie Stepenoff, "The Last Tree Cut Down: The End of the Bootheel Frontier, 1880–1940," *Missouri Historical Review*, 90(October 1995): 68.

9. H. A. Kipp, *Report on the Belzoni Drainage District in Washington County, Mississippi*, U.S. Department of Agriculture, Office of Experiment Stations, Bulletin No. 244 (Washington, D.C.: Government Printing Office, 1912). 10–14; U.S. Congress, House of Representatives, *Yazoo River, Mississippi*, 67th Congress, 4th Session, House Document No. 597 (Washington, D.C.: Government Printing Office, 1923), 15; Marsden and Teele, *Economic Status of Drainage Districts*, 5.

10. Jeannie Whayne, "Creation of a Plantation System in the Twentieth Century," *Agricultural History*, 66(Winter 1992): 63–83; Jeannie Whayne, *A New Plantation South* (Charlottesville, Va.: University of Virginia Press, 1996), 104–111.

11. Clifton Johnson, *Highways and Byways of the Mississippi Valley* (New York, N.Y.: MacMillan Co., 1906), 111–114.

12. Marsden and Teele, *Economic Status of Drainage Districts*, 29–30, 33. See also C. Ray Brassieur, "From Two Mules to Twelve-Row Equipment: An Oral His-

tory Interview with Maxwell Williams, Bootheel Farm Manager," *Missouri Historical Review*, 91(October 1996): 59.

13. Marsden and Teele, *Economic Status of Drainage Districts*, 36, 43, 45–47.

14. Marsden and Teele, *Economic Status of Drainage Districts*, 10–17, 28.

15. Charles W. Okey, *The Wet Lands of Southern Louisiana and Their Drainage*, U.S. Department of Agriculture, Department Bulletin No. 652 (Washington, D.C.: Government Printing Office, 1918), 1–15; Harrison, *Alluvial Empire*, Vol. 1:208–210, 252–272.

16. Morgan and others, *Drainage of the Fifth Louisiana Levee District*, 5–29.

17. Robert W. Harrison and Walter M. Kollmorgen, *Drainage Reclamations in the Bartholomew–Boeuf–Tensas Basin of Arkansas and Louisiana*, Arkansas Agricultural Experiment Station, Bulletin No. 476 (Fayetteville: University of Arkansas, 1948), 3–10, 45–55; R. D. Calhoun, "A History of Concordia Parish, Louisiana," II, *Louisiana Historical Quarterly*, 15(April 1932): 224.

18. Marsden and Teele, *Economic Status of Drainage Districts*, 29–30.

19. Dan A. Rudd and Theo. Bond, *From Slavery to Wealth: The Life of Scott Bond* (Madison, Ark.: Journal Printing Co., 1917), 365–366.

20. See Alfred H. Stone, "The Negro Farmer in the Mississippi Delta," *Southern Workman*, 32(October 1903): 458; Harrison, *Alluvial Empire*, Vol. 1:303.

21. Carl Kelsey, *The Negro Farmer* (Chicago, Ill.: Jennings and Pye, 1903), 53.

22. Thad Snow, *From Missouri* (Boston, Mass.: Houghton Mifflin Co., 1954), 2, 133–134.

23. Snow, *From Missouri*, 135.

24. Melinda S. Meade, "The Rise and Demise of Malaria," *Southeastern Geographer*, 20(November 1980): 77–78, 86.

25. *The West-Side Delta* (Memphis, Tenn.: Southern Alluvial Land Association, 1920), 41.

26. Johnson, *Highways and Byways*, 122; Meade, "Rise and Demise of Malaria," 86.

27. Snow, *From Missouri*, 136–137.

28. *The Yazoo–Mississippi Delta* (Memphis, Tenn.: Southern Alluvial Land Association, 1920), 27; William F. Gray, ed., *Imperial Bolivar* (Cleveland, Miss.: Bolivar Commercial, 1923), 38.

29. *The West-Side Delta*, 43; Roy Swank, *Trail to Marked Tree* ed. by Nolan Porterfield (San Antonio, Tex.: Naylor Co., 1968), 95–96, 98–99.

30. Max R. White and others, *Rich Land—Poor People*, Farm Security Administration, Research Report No. 1 (Indianapolis, Ind.: Farm Security Administration, Region III, 1938), 5–6, 44, 47–48; Meade, "Rise and Demise of Malaria," 87, 89.

31. A. C. True, prep., *The Cotton Plant*, U.S. Department of Agriculture, Office of Experiment Stations, Bulletin No. 33 (Washington, D.C.: Government Printing Office, 1896), 255; S. M. Tracy, *Corn Culture in the South*, U.S. Department of Agriculture, Farmers' Bulletin No. 81 (Washington, D.C.: Government Printing Office, 1898), 6–7.

32. L. O. Howard, *Insects Affecting the Cotton Plant*, U.S. Department of Agriculture, Farmers' Bulletin No. 47 (Washington, D.C.: Government Printing Office, 1897), 3, 8–12, 16–21, 24–25; Rudd and Bond, *From Slavery to Wealth*, 331.

33. W. D. Hunter, "Present Status of the Mexican Boll-Weevil in the United

States," in U.S. Department of Agriculture, *Yearbook of the U.S. Department of Agriculture: 1901* (Washington, D.C.: Government Printing Office, 1902), 369–380.

34. W. D. Hunter, "The Status of the Mexican Boll-Weevil in the United States in 1903," in U.S. Department of Agriculture, *Yearbook of the U.S. Department of Agriculture: 1903* (Washington, D.C.: Government Printing Office, 1904), 205–207; W. D. Hunter and B. R. Coad, *The Boll-Weevil Problem*, U.S. Department of Agriculture, Farmers' Bulletin No. 1262 (Washington, D.C.: Government Printing Office, 1922), 5–6; Kent Osband, "The Boll Weevil Versus 'King Cotton,' " *Journal of Economic History*, 45(September 1985): 627.

35. W. D. Hunter, *The Control of the Boll-Weevil*, U.S. Department of Agriculture, Farmers' Bulletin No. 216 (Washington, D.C.: Government Printing Office, 1905), 3–5; W. D. Hunter, "Some Recent Studies of the Mexican Cotton Boll-Weevil," in U.S. Department of Agriculture, *Yearbook of the U.S. Department of Agriculture: 1906* (Washington, D.C.: Government Printing Office, 1907), 322–323; O. F. Cook, *Cotton Improvement Under Weevil Conditions*, U.S. Department of Agriculture, Farmers' Bulletin No. 501 (Washington, D.C.: Government Printing Office, 1912), 7–17.

36. Hunter, "Some Recent Studies," 315–316.

37. R. D. Calhoun, "A History of Concordia Parish, Louisiana," VII, *Louisiana Historical Quarterly*, 16(October 1933): 604–605; Frederick W. Williamson and George T. Goodman, eds., *Eastern Louisiana* (Monroe, La.: Historical Record Association, 1939), Vol. 2:762

38. William L. Giles, "Agricultural Revolution, 1890–1970," in Richard A. McLemore, ed., *A History of Mississippi* (Jackson, Miss.: University Press of Mississippi, 1973), Vol. 2:191–194.

39. W. D. Hunter, "Relation Between Rotation Systems and Insect Injury in the South," in U.S. Department of Agriculture, *Yearbook of the U.S. Department of Agriculture: 1911* (Washington, D.C.: Government Printing Office, 1912), 202–203, 207–208.

40. Brandfon, *Cotton Kingdom of the New South*, 123.

41. B. R. Coad and T. P. Cassidy, *Cotton Boll-Weevil Control by the Use of Poison*, U.S. Department of Agriculture, Department Bulletin No. 875 (Washington, D.C.: Government Printing Office, 1920), 3–4, 8–29; Douglas Helms, "Technological Methods for Boll Weevil Control," *Agricultural History*, 53(January 1979): 291–292.

42. W. D. Hunter, *The Boll-Weevil Problem*, U.S. Department of Agriculture, Farmers' Bulletin No. 848 (Washington, D.C.: Government Printing Office, 1917), 23; A. D. McNair, *Crop Systems for Arkansas*, U.S. Department of Agriculture, Farmers' Bulletin No. 1000 (Washington, D.C.: Government Printing Office, 1918), 5; Hunter and Coad, "Boll Weevil Problem," 5.

43. Lewis E. Hutchins, comp., *Resources of Arkansas* (Little Rock, Ark.: State Bureau of Mines, Manufactures and Agriculture, 1911), 9–10, 14–16.

44. *A History of the North-Western Editorial Excursion to Arkansas* (Little Rock, Ark.: T. B. Mills, 1876), 255; W. A. Browne, "Grand Prairie: A Progressive Rice Region," *Journal of Geography*, 31(April 1932): 138, 146; Ernest E. Sampson, "Half a Century on Grand Prairie," *Arkansas Historical Quarterly*, 14(Spring 1955): 32–33, 35.

45. See *Experiment Station Work, XLII*, U.S. Department of Agriculture, Far-

mers' Bulletin No. 305 (Washington, D.C.: Government Printing Office, 1907), 5–6; Florence L. Rosencrantz, "The Rice Industry in Arkansas," *Arkansas Historical Quarterly*, 5(Summer 1946): 123–124; Henry C. Dethloff, "Rice Revolution in the Southwest," *Arkansas Historical Quarterly*, 29(Spring 1970): 72–73; Pete Daniel, *Breaking the Land: The Transformation of Cotton, Tobacco and Rice Culture Since 1880* (Urbana, Ill.: University of Illinois Press, 1985), 39–49.

46. *Arkansas* (Washington, D.C.: U.S. Railroad Administration, 1919), 15–17; Thomas Foti, "The River's Gifts and Curses," in Jeannie Whayne and Willard Gatewood, eds., *The Arkansas Delta* (Fayetteville, Ark.: University of Arkansas Press, 1993), 44–46.

47. Daniel, *Breaking the Land*, 49; Henry C. Dethloff, *A History of the American Rice Industry, 1685–1985* (College Station, Tex.: Texas A & M University Press, 1988), 81, 86–87.

48. A. D. McNair, *Labor Requirements of Arkansas Crops*, U.S. Department of Agriculture, Department Bulletin No. 1181 (Washington, D.C.: Government Printing Office, 1924), 61.

49. *The Arkansas Handbook* (Ft. Smith, Ark.: State Bureau of Mines, Manufactures and Agriculture, 1924), 46.

50. Dethloff, "Rice Revolution in the Southwest," 74; Dethloff, *American Rice Industry*, 85–87.

51. Hutchins, comp., *Resources of Arkansas, 9; Arkansas*, 15–17.

52. George F. Jenks and Robert W. Harrison, *Methods and Costs of Clearing Land in Northeast Arkansas*, Arkansas Agricultural Experiment Station, Bulletin No. 495 (Fayetteville, Ark.: University of Arkansas, 1950), 3, 10, 12–13, 16; Robert W. Harrison, "Clearing Land in the Mississippi Alluvial Valley," *Arkansas Historical Quarterly*, 13(Winter 1954): 354–356, 365, 367; William M. Clements and Larry D. Ball, "This Was the Beginning of Clearing of Land: The Development and Use of the East Arkansas Stump Saw," *Arkansas Historical Quarterly*, 45(Spring 1986): 42–46.

53. *Arkansas*, 16; Samuel T. Bratton, "The Geography of the St. Francis Basin," *University of Missouri Studies*, 1(July 1926): 23–26.

54. Hunter and Coad, "Boll-Weevil Problem," 5; Bratton, "Geography of the St. Francis Basin," 48; Lee A. Dew, "The J.L.C. and E.R.R. and the Opening of the 'Sunk Lands' of Northeast Arkansas," *Arkansas Historical Quarterly*, 27(Spring 1968): 30–32; Whayne, "Creation of a Plantation System," 76–77; Whayne, *New Plantation South*, 113–116.

55. Bratton, "Geography of the St. Francis Basin," 26–27; *The West-Side Delta*, 9, 56, 63.

56. U.S. Census Office, *Tenth Census (1880): Report on Cotton Production in the United States* (Washington, D.C.: Government Printing Office, 1884), Part 1: 549–550.

57. *Arkansas*, 4, 38; Bratton, "Geography of the St. Francis Basin," 8; Dew, "The J.L.C. and E.R.R.," 29–32; Whayne, *New Plantation South*, 23–24, 99–112.

58. See Rudd and Bond, *From Slavery to Wealth*, 348–349; *The West-Side Delta*, 9, 56; Mabel F. Edrington, *History of Mississippi County, Arkansas* (Ocala Fla.: Ocala Star Banner, 1962), 65–66, 319–320; Dew, "The J.L.C. and E.R.R.," 28.

59. Dew, "The J.L.C. and E.R.R." 31; Swank, *Trail to Marked Tree*, 5, 17.

60. Bratton, "Geography of the St. Francis Basin," 1, 26.

61. *Arkansas*, 3.

62. *Missouri* (Washington, D.C.: U.S. Railroad Administration, 1919), 8, 11–13, 24, 33; Bratton, "Geography of the St. Francis," 26.

63. Snow, *From Missouri*, 2.

64. Bratton, "Geography of the St. Francis Basin," 23.

65. Hugh D. Studabaker, *What They "Showed Me" in Southeast-Missouri* (privately printed, 1913), 3, 6–7, 28, 31.

66. Snow, *From Missouri*, 6, 8, 87, 90–97, 110, 124.

67. Bratton, "Geography of the St. Francis Basin," 43; White and others, *Rich Land—Poor People*, 4.

68. Stepenoff, "Last Tree Cut Down," 71–72.

69. H. R. Cates, *Farm Practice in the Cultivation of Cotton*, U.S. Department of Agriculture, Department Bulletin No. 511 (Washington, D.C.: Government Printing Office, 1917), 4, 17–20.

70. Lewis F. Thomas, "Basic Factors in Flood Frequency in the Lower Mississippi River," *Annals of the American Academy of Political and Social Science*, 135(January 1928): 1; Harrison, *Alluvial Empire*, Vol. 1:129.

71. Snow, *From Missouri*, 98–108.

72. S. E. Simonson, "The St. Francis Levee and High Water on the Mississippi River," *Arkansas Historical Quarterly*, 6(Winter 1947): 422–427.

73. R. D. Calhoun, "A History of Concordia Parish," IV, *Louisiana Historical Quarterly*, 15(October 1932): 630.

74. Harrison, *Levee Districts and Levee Building in Mississippi*, 188–218; Harrison, *Alluvial Empire*, Vol. 1:125, 129–130.

75. *The Yazoo–Mississippi Delta*, 31; Harrison, *Levee Districts and Levee Building in Mississippi*, 226–231; Harrison, *Alluvial Empire*, Vol. 1:131–134; Arthur D. Frank, *The Development of the Federal Program of Flood Control on the Mississippi River* (New York, N.Y.: AMS Press, 1968), 168–171.

76. Gray, ed., *Imperial Bolivar*, 29.

77. Frederick Strauss and Louis Bean, *Gross Farm Income and Indices of Farm Production and Prices in the United States, 1869–1937*, U.S. Department of Agriculture, Technical Bulletin No. 703 (Washington, D.C.: Government Printing Office, 1940), 64–66; Brandfon, *Cotton Kingdom of the New South*, 128–130, 135; Cobb, *Most Southern Place on Earth*, 100.

78. See Paul F. George, "Welfare Work on a Delta Plantation," *Southern Workman*, 54(July 1925): 314–315; "Biggest Cotton Plantation," *Fortune*, 15(March 1937): 127–132, 156–160; Oscar Johnston, "Industrial Plan of the Delta and PineLand Company of Mississippi," *Journal of Social Hygiene*, 26(February 1940): 73–74; Florence W. Sillers, comp., *History of Bolivar County, Mississippi* ed. by Wirt A. Williams (Jackson, Miss.: Hederman Bros., 1948), 250–253.

79. E. A. Boeger and E. A. Goldenweiser, *A Study of the Tenant Systems of Farming in the Yazoo–Mississippi Delta*, U.S. Department of Agriculture, Department Bulletin No. 337 (Washington, D.C.: Government Printing Office, 1916), 3, 6–7.

Between 1900 and 1920, the proportion of renter-operated farms in the Mississippi Delta decreased from 44 percent to 19 percent, whereas the cropper-operated farms increased from 46 percent to 74 percent. See John C. Willis, "On the New South Frontier: Life in the Yazoo–Mississippi Delta, 1865–1920" (Ph.D. dissertation, University of Virginia, Charlottesville, 1991), 357–358.

80. Cates, *Farm Practice in the Cultivation of Cotton*, 21–23.

81. Boeger and Goldenweiser, *Tenant Systems of Farming*, 1, 6–7; Cobb, *Most Southern Place on Earth*, 101–104, 112–113.

82. Gray, ed., *Imperial Bolivar*, 18–19.

83. George W. Hill, "The Bureau of Animal Industry," *Chautauquan*, 14(February 1892): 549–551.

84. John R. Mohler, *Texas or Tick Fever and Its Prevention*, U.S. Department of Agriculture, Farmers' Bulletin No. 258 (Washington, D.C.: Government Printing Office, 1906), 5–8, 32, 39, 41, 45; W. P. Ellenberger and Robert M. Chapin, *Cattle-Fever Ticks and Methods of Eradication*, U.S. Department of Agriculture, Farmers' Bulletin No. 1057 (Washington, D.C.: Government Printing Office, 1919), 3–5, 11–13, 23–24; *The Yazoo–Mississippi Delta*, 23; Frederick W. Williamson, *Origin and Growth of Agricultural Extension in Louisiana, 1860–1948* (Baton Rouge, La.: Louisiana State University, 1951), 148–149.

85. M. Dorset, *Hog Cholera*, U.S. Department of Agriculture, Farmers' Bulletin No. 379 (Washington, D.C.: Government Printing Office, 1909), 5–7, 19; *The West-Side Delta*, 25.

86. F. W. Farley, "Growth of the Beef-Cattle Industry in the South," in U.S. Department of Agriculture, *Yearbook of the U.S. Department of Agriculture: 1917* (Washington, D.C.: Government Printing Office, 1918), 327.

87. S. M. Tracy, *Forage for the Cotton Belt*, U.S. Department of Agriculture, Farmers' Bulletin No. 1125 (Washington, D.C.: Government Printing Office, 1920), 51–58.

88. W. F. Ward and Dan T. Gray, *Beef Production in the South*, U.S. Department of Agriculture, Farmers' Bulletin No. 580 (Washington, D.C.: Government Printing Office, 1914), 3–4.

89. Lyman Carrier and F. G. Ashbrook, *Hog Pastures for the Southern States*, U.S. Department of Agriculture, Farmers' Bulletin No. 951 (Washington, D.C.: Government Printing Office, 1918), 5–7, 9–10, 13.

90. See Ward and Gray, *Beef Production in the South*, 5, 15–18; Farley, "Growth of the Beef Cattle Industry," 327–331, 340; *The Yazoo–Mississippi Delta, 25; The West-Side Delta*, 49; E. Z. Russell, *Breeds of Swine*, U.S. Department of Agriculture, Farmers' Bulletin No. 1263 (Washington, D.C.: Government Printing Office, 1922), 4–5, 8.

91. *The Yazoo–Mississippi Delta*, 45, 47.

92. *The West-Side Delta*, 49; Bratton, "Geography of the St. Francis Basin," 51.

93. Strauss and Bean, *Gross Farm Income*, 110, 119.

CHAPTER 4

1. U.S. Congress, House of Representatives, *Reclamation and Rural Development in the South*, 69th Congress, 2nd Session, House Document No. 765 (Washington, D.C.: Government Printing Office, 1927), Part 2: 78.

2. Frederick Strauss and Louis Bean, *Gross Farm Income and Indices of Farm Production and Prices in the United States*, 1869–1937, U.S. Department of Agriculture, Technical Bulletin No. 703 (Washington, D.C.: Government Printing Office, 1940), 65–66; William F. Gray, ed., *Imperial Bolivar* (Cleveland, Miss.: Bolivar Commercial, 1923), 19.

3. Howard Snyder, "Negro Migration and the Cotton Crop," *North American Review*, 219(January 1924): 21–22.

4. U.S. Congress, *Reclamation and Rural Development*, Part 2: 46–47.

5. See B. R. Coad, "Airplane Dusting of Cotton Fields Proves Effective, Economical," in U.S. Department of Agriculture, *Yearbook of Agriculture: 1928* (Washington, D.C.: Government Printing Office, 1929), 117, 119; Florence W. Sillers, comp., *History of Bolivar County, Mississippi* ed. by Wirt A. Williams (Jackson, Miss.: Hederman Bros., 19480, 250; Eldon W. Downs and George F. Lemmer, "Origins of Aerial Crop Dusting," *Agricultural History*, 39(July 1965): 133; Douglas Helms, "Technological Methods for Boll-Weevil Control," *Agricultural History*, 53(January 1979): 291.

6. Elmer Johnson and B. R. Coad, *Dusting Machinery for Cotton Boll-Weevil Control*, U.S. Department of Agriculture, Farmers' Bulletin No. 1098 (Washington, D.C.: Government Printing Office, 1920), 7–9, 12–18; W. D. Hunter and B. R. Coad, *The Boll-Weevil Problem*, U.S. Department of Agriculture, Farmers' Bulletin No. 1262 (Washington, D.C.: Government Printing Office, 1922), 16–18; George R. Boyd, *Clearing Land of Brush and Stumps*, U.S. Department of Agriculture, Farmers' Bulletin No. 1526 (Washington, D.C.: Government Printing Office, 1927), 13.

7. See H. R. Tolley and L. M. Church, *Tractors on Southern Farms*, U.S. Department of Agriculture, Farmers' Bulletin No. 1278 (Washington, D.C.: Government Printing Office, 1922), 3; Robert C. Williams, *Fordson, Farmall and Poppin' Johnny: A History of the Farm Tractor and Its Impact on America* (Urbana, Ill.: University of Illinois Press, 1987), 48–49; R. D. Hurt, "Agricultural Technology in the Twentieth Century," *Journal of the West*, 30(April 1991): 18–19.

8. Tolley and Church, *Tractors on Southern Farms*, 6–8; Jack T. Kirby, *Rural Worlds Lost: The American South*, 1920–1960 (Baton Rouge, La.: Louisiana State University Press, 1987), 34, 54–55, 197–200.

9. Sydney Nathans, "Gotta Mind to Move, A Mind to Settle Down: Afro-Americans and the Plantation Frontier," in William J. Cooper, Jr., and others, eds., *A Master's Due: Essays in Honor of David Herbert Donald* (Baton Rouge, La.: Louisiana State University Press, 1985), 216.

10. See William L. Giles, "Agricultural Revolution, 1890–1970," in Richard A. McLemore, ed., *A History of Mississippi* (Jackson, Miss.: University Press of Mississippi, 1973), Vol. 2: 197; James C. Cobb, *Most Southern Place on Earth* (New York, N.Y.: Oxford University Press, 1992), 116–120, 124.

11. U.S. Congress, House of Representatives, *Yazoo River, Mississippi*, 67th Congress, 4th Session, House Document No. 597 (Washington, D.C.: Government Printing Office, 1923), 29; U.S. Congress, *Reclamation and Rural Development*, Part 2: 51–52; Boyd, *Clearing Land of Brush and Stumps*, 9, 13.

12. Gray, ed., *Imperial Bolivar*, 22–23; Sillers, comp., *History of Bolivar County*, 254–257.

13. M. G. Vaiden and others, *Making Cotton Cheaper: Can Present Production Costs Be Reduced?*, Mississippi Agricultural Experiment Station, Bulletin No. 290 (Starkville, Miss.: Mississippi A & M College, 1931), 4.

14. Fern E. Dorris, "The Yazoo Basin in Mississippi," *Journal of Geography*, 28(February 1929): 78.

15. C. A. McCombs, "The Present Status of Navigation on the Lower Missis-

sippi River," *Journal of Geography*, 24(January 1925):12; Harry P. Owens, *Steamboats and the Cotton Economy* (Jackson, Miss.: University Press of Mississippi, 1990), 150, 152, 154.

16. E. L. Langsford and B. H. Thibodeaux, *Plantation Organization and Operation in the Yazoo–Mississippi Delta Area*, U.S. Department of Agriculture, Technical Bulletin No. 682 (Washington, D.C.: Government Printing Office, 1939), 13.

17. Gray, ed., *Imperial Bolivar*, 39–40; Dunbar Rowland, *History of Mississippi* (Chicago, Ill: S. J. Clarke, 1925), Vol. 2: 558–559; Robert L. Brandfon, *Cotton Kingdom of the New South* (Cambridge, Mass.: Harvard University Press, 1967), 83–84.

18. Gray, ed., *Imperial Bolivar*, 17.

19. Maurice O. Eldridge, *Good Roads for Farmers*, U.S. Department of Agriculture, Farmers' Bulletin No. 95 (Washington, D.C.: Government Printing Office, 1899), 7–8, 13, 27, 29, 39; Joseph H. Pratt, "Good Roads Movement in the South," *Annals of the American Academy of Political and Social Science*, 35(January–June 1910): 109; J. E. Pennybaker, "State Management of Public Roads," in U.S. Department of Agriculture, *Yearbook of the U.S. Department of Agriculture: 1914* (Washington, D.C.: Government Printing Office, 1915), 212; Thomas D. Clark, "Changes in Transportation," in McLemore, ed., *A History of Mississippi*, Vol. 2: 285.

20. T. Warren Allen, "Highways and Highway Transportation," in U.S. Department of Agriculture, *Agriculture Yearbook: 1924* (Washington, D.C.: Government Printing Office, 1925), 100–101, 104, 108–109, 111–113, 115, 119, 121–127; Marsha P. Hataway, "The Development of the Mississippi State Highway System, 1916–1932,"*Journal of Mississippi History,* 28(November 1966): 286–287, 289–290, 293–295; Clark, "Changes in Transportation," Vol. 2:287–290; Howard Preston, *Dirt Roads to Dixie: Accessibility and Modernization in the South, 1885–1935* (Knoxville, Tenn.: University of Tennessee Press, 1991), 154–158.

21. *The Yazoo–Mississippi Delta* (Memphis, Tenn.: Southern Alluvial Land Association, 1920), 37, 39; Gray, ed., *Imperial Bolivar*, 31; Allen, "Highways and Highway Transportation," 126, 141, 167; U.S. Congress, *Reclamation and Rural Development*, Vol. 2:59.

22. U.S. Congress, *Yazoo River, Mississippi*, 1, 4; U.S. Congress, *Reclamation and Rural Development*, Part 2:39, 57; Walter Sillers, Sr., "Flood Control in Bolivar County, 1838–1924," *Journal of Mississippi History*, 9(January 1947): 19–20; Robert W. Harrison, *Levee Districts and Levee Building in Mississippi* (Stoneville, Miss.: Delta Council and others, 1951), 227–229, 231.

23. *The West-Side Delta* (Memphis, Tenn.: Southern Alluvial Land Association, 1920), 17, 31, 35; R. D. Calhoun, "A History of Concordia Parish, Louisiana," II, VII, and VIII, *Louisiana Historical Quarterly*, 15(April 1932): 224; 16(July 1933): 472; and 16(October 1933): 604–605; Frederick W. Williamson and George T. Goodman, eds., *Eastern Louisiana* (Monroe, La.: Historical Record Association, 1939), Vol. 1: 192, 253.

24. *The West-Side Delta*, 32–33.

25. Stephen E. Wood, "The Development of Arkansas Railroads: Part II," *Arkansas Historical Quarterly*, 7(Autumn 1948): 178, 180; Lee A. Dew, "The J.L.C. and E.R.R. and the Opening of the 'Sunk Lands' of Northeast Arkansas," *Arkansas Historical Quarterly*, 27(Spring 1968): 23–30, 33, 39; Clifton E. Hull, *Shortline*

Railroads of Arkansas (Norman, Okl.: University of Oklahoma Press, 1969), 51–55, 72–73, 93, 384, 386; Lawrence R. Handley, "Settlement Across Northern Arkansas as Influenced by the Missouri & North Arkansas Railroad," *Arkansas Historical Quarterly, 33(Winter 1974): 273, 281.*

26. *Arkansas Facts* (Little Rock, Ark.: State Bureau of Mines, Manufactures and Agriculture, 1924), 29–30; Wood, "Development of Arkansas Railroads: Part II," 176, 178, 183, 185, 187, 189–190; Ira G. Clark, *Then Came the Railroads* (Norman, Okl.: University of Oklahoma Press, 1958), 236, 238–239; Hull, *Shortline Railroads of Arkansas*, 384.

27. *The Arkansas Handbook* (Ft. Smith, Ark.: State Bureau of Mines, Manufactures and Agriculture, 1924), 32.

28. *The West-Side Delta*, 37–39, 53; Samuel T. Bratton, "The Geography of the St. Francis Basin," *University of Missouri Studies*, 1(July 1926): 20, 22; Dew, "The J.L.C. and E.R.R.," 37–39; Jeannie Whayne, "Creation of a Plantation System in the Arkansas Delta in the Twentieth Century," *Agricultural History*, 66(Winter 1992): 68–70.

29. Strauss and Bean, *Gross Farm Income*, 65–66.

30. *Economic Survey of Arkansas* (St. Louis, Mo.: Southwestern Bell Telephone Co., 1925), 106, 117; Jeannie M. Whayne, *A New Plantation South* (Charlottesville, Va.: University Press of Virginia, 1996), 140–141.

31. Donald C. Alexander, *The Arkansas Plantation, 1920–1942* (New Haven, Conn.: Yale University Press, 1943), 11–12, 16.

32. Hunter and Coad, *Boll Weevil Problem*, 5–6; Bratton, "Geography of the St. Francis Basin," 48.

33. Roy Swank, *Trail to Marked Tree* ed. by Nolan Poterfield (San Antonio, Tex.: Naylor Co. 1968), 17, 46–51, 53–54.

34. Strauss and Bean, *Gross Farm Income*, 71; Florence L. Rosencrantz, "The Rice Industry in Arkansas," *Arkansas Historical Quarterly*, 5(Summer 1946): 128.

35. Pete Daniel, *Breaking the Land* (Urbana, Ill.: University of Illinois Press, 1985), 58; Whayne, *New Plantation South*, 122.

36. George F. Jenks and Robert W. Harrison, *Methods and Costs of Clearing Land in Northeast Arkansas*, Arkansas Agricultural Experiment Station, Bulletin No. 495 (Fayetteville, Ark.: University of Arkansas, 1950), 4–5.

37. Bratton, "Geography of the St. Francis Basin," 9–10, 20, 22, 48; Leon P. Ogilvie, "The Development of the Southeast Missouri Lowlands" (Ph.D. dissertation, University of Missouri, Columbia, 1967), 65–66.

38. Bratton, "Geography of the St. Francis Basin," 23, 26–28.

39. Bratton, "Geography of the St. Francis Basin," 28.

40. Bratton, "Geography of the St. Francis Basin," 32, 34–35, 38; Oliver E. Baker, "Agricultural Regions of North America: Part I: The Basis of Classification," *Economic Geography*, 3(January 1926): 473; Ogilvie, "Development of the Southeast Missouri Lowlands," 178–189.

41. Bratton, "Geography of the St. Francis Basin," 37–38, 43, 50; Max R. White and others, *Rich Land—Poor People*, Farm Security Administration, Research Report No. 1 (Indianapolis, Ind.: Farm Security Administration, Region III, 1938), 4; Bonnie Stepenoff, "The Last Tree Cut Down: The End of the Bootheel Frontier, 1880–1940," *Missouri Historical Review*, 90(October 1995): 72.

42. Thad Snow, *From Missouri* (Boston, Mass.: Houghton Mifflin Co., 1954), 157.

43. White and others, *Rich Land—Poor People*, 4, 19; Irvin G. Wyllie, "Race and Class Conflict on Missouri's Cotton Frontier," *Journal of Southern History*, 20(May 1954): 195; Ogilvie, "Development of the Southeast Missouri Lowlands," 252– 260; Stepenoff, "Last Tree Cut Down," 73–74.

44. Snow, *From Missouri*, 157–158, 160.

45. Bratton, "Geography of the St. Francis Basin," vi, 12, 15–16; Oliver E. Baker, "Agricultural Regions of North America: Part II: The South," *Economic Geography*, 3(January 1927): 65 and "Agricultural Regions of North America: Part IV: The Corn Belt," *Economic Geography*, 3(October 1927): 447; White and others, *Rich Land—Poor People*, 5–10.

46. Andrew G. Chilton, *Through Arkansas on the Hog* (Chicago, Ill.: M. A. Donohue and Co., 1908), 10.

47. J. Russell Smith, *North America* (New York, N.Y.: Harcourt, Brace and Co., 1925), 248.

48. Ernest C. Faust, "The History of Malaria in the United States," *American Scientist*, 39(January 1951): 123; Bob Lancaster, *The Jungles of Arkansas: A Personal History of the Wonder State* (Fayetteville, Ark.: University of Arkansas Press, 1989), 99.

49. See Faust, "History of Malaria," 125; Melinda S. Meade, "The Rise and Demise of Malaria," *Southeastern Geographer*, 20(November 1980): 89–90; Lancaster, *Jungles of Arkansas*, 100–101.

50. S. E. Simonson, "Origin of Drainage Projects in Mississippi County," *Arkansas Historical Quarterly*, 5(Fall 1946): 273; Brandfon, *Cotton Kingdom of the New South*, 126.

51. *The Mississippi Valley Flood Disaster of 1927: Official Report of the Relief Operations* (Washington, D.C.: American National Red Cross, [1929]), 2, 4; John M. Barry, *Rising Tide: The Great Mississippi Flood of 1927 and How It Changed America* (New York, N.Y.: Simon and Schuster, 1997), 173–188.

52. F. H. Newell, "Mississippi River Floods," *American Review of Reviews*, 75(June 1927): 593–594; Barry, *Rising Tide*, 194.

53. Arthur Kellogg, "Behind the Levees," *Survey*, 58(June 1, 1927): 281; *Mississippi Valley Flood Disaster*, 6; James H. Street, *Look Away! A Dixie Notebook* (Westport, Conn.: Greenwood Press, 1977), 168–169.

54. Thomas Fauntleroy, "Lord, Plant My Feet on Higher Ground: The Mississippi at Its Worst," *Independent*, 118(June 4, 1927): 581; Kellogg, "Behind the Levees," 282; Sillers, comp., *History of Bolivar County*, 193, 203; Barry, *Rising Tide*, 201–208.

55. Kellogg, "Behind the Levees," 282; "Four Stages of the Flood," *Survey*, 58(July 1, 1927): 357; Barry, *Rising Tide*, 240, 262, 272–273.

56. Helen Murphy, "Overflow Notes," *Atlantic Monthly*, 140(August 1927): 223–224.

57. Murphy, "Overflow Notes," 226–229; Barry, *Rising Tide*, 281.

58. Will Irwin, "Can We Tame the Mississippi?" *World's Work*, 54(August 1927): 407; *Mississippi Valley Flood Disaster*, 56; William A. Percy, *Lanterns on the Levee: Recollections of a Planter's Son* (New York, N.Y.: A. A. Knopf, 1941), 249; Sillers, comp., *History of Bolivar County*, 203, 250; Barry, *Rising Tide*, 284, 324.

59. J. E. Switzer, "A Trip to the Mississippi–Yazoo Flood District," *Journal of Geography*, 26(November 1927): 293–298; *Mississippi Valley Flood Disaster*, 122–126; R. D. Calhoun, "A History of Concordia Parish, Louisiana," IV, *Louisiana Historical Quarterly*, 15(October 1932): 630; Bern Keating, *A History of Washington County, Mississippi* (Greenville, Miss.: Greenville Junior Auxiliary, 1976), 65–67.

60. "Biggest Cotton Plantation," *Fortune*, 15(March 1937): 130–131; Sillers, ed., *History of Bolivar County*, 250–251.

61. "Biggest Cotton Plantation," 131; Barry, *Rising Tide*, 313–314.

62. Walter White, "The Negro and the Flood," *Nation*, 124(June 22, 1927): 688–689.

63. Arthur Kellogg, "Up From the Bottom Lands," *Survey*, 58(July 1, 1927): 363–364; *Mississippi Valley Flood Disaster*, 67–68.

64. Street, *Look Away!*, 176.

65. *Mississippi Valley Flood Disaster*, 64, 107; Barry, *Rising Tide*, 387.

66. Cobb, *Most Southern Place on Earth*, 122–123; Barry, *Rising Tide*, 416–417.

67. Barry, *Rising Tide*, 387.

68. See Kellogg, "Behind the Levees," 290 and "Up From the Bottom Lands," 361, 363; *Mississippi Valley Flood Disaster*, 65, 89–90; Whayne, *New Plantation South*, 147.

69. *Mississippi Valley Flood Disaster*, 68–69.

70. Kellogg, "Up From the Bottom Lands," 360.

71. *Mississippi Valley Flood Disaster*, 65, 67–68, 132.

72. W. J. Morse, *The Soy Bean: Its Culture and Uses*, U.S. Department of Agriculture, Farmers' Bulletin No. 973 (Washington, D.C.: Government Printing Office, 1918), 3, 15, 31; A. G. Smith, *Soy Beans in Systems of Farming in the Cotton Belt*, U.S. Department of Agriculture, Farmers' Bulletin No. 931 (Washington, D.C.: Government Printing Office, 1918), 3, 6, 16; S. M. Tracy, *Forage for the Cotton Belt*, U.S. Department of Agriculture, Farmers' Bulletin No. 1125 (Washington, D.C.: Government Printing Office, 1920), 37–38; *The Yazoo-Mississippi Delta*, 19.

73. C. V. Piper and H. T. Nielsen, *Soy Beans*, U.S. Department of Agriculture, Farmers' Bulletin No. 372 (Washington, D.C.: Government Printing Office, 1909), 15–17; Smith, *Soy Beans in Systems of Farming*, 8–9; Tracy, *Forage for the Cotton Belt*, 38; A. D. McNair, *Crop Systems for Arkansas*, U.S. Department of Agriculture, Farmers' Bulletin No. 1000 (Washington, D.C.: Government Printing Office, 1918), 19; P. G. Bedenbaugh, *Grazing and Feeding Trials with Corn and Soybeans for Pork Production*, Mississippi Agricultural Experiment Station, Bulletin No. 283 (Starkville, Miss.: Mississippi A & M College, 1930), 4–5.

74. Jared G. Smith, *Alafalfa or Lucern*, U.S. Department of Agriculture, Farmers' Bulletin No. 31 (Washington, D.C.: Government Printing Office, 1899), 14–20; *The Yazoo–Mississippi Delta*, 19; Tracy, *Forage for the Cotton Belt*, 29–31; A. D. McNair, *Labor Requirements of Arkansas Crops*, U.S. Department of Agriculture, Department Bulletin No. 1181 (Washington, D.C.: Government Printing Office, 1924), 25.

75. A. D. McNair and W. B. Mercie, *Lespedeza or Japan Clover*, U.S. Department of Agriculture, Farmers' Bulletin No. 441 (Washington, D.C.: Government Printing Office, 1911), 5–9; Lyman Carrier, *Lespedeza as a Forage Crop*, U.S. Department of Agriculture, Farmers' Bulletin No. 1143 (Washington, D.C.: Government Printing Office, 1920), 3–4, 8–12, 14; Sillers, comp., *History of Bolivar County*, 251.

76. Tracy, *Forage for the Cotton Belt*, 34.

77. *Mississippi Valley Flood Disaster*, 70–73.

78. Strauss and Bean, *Gross Farm Income*, 110, 119; U.S. Census Bureau, *Fifteenth Census of the United States: 1930: Agriculture* (Washington, D.C.: Government Printing Office, 1932), Part 2: 1237.

79. Calhoun, "History of Concordia Parish," VII, 606.

80. Kellogg, "Up From the Bottom Lands," 366; Strauss and Bean, *Gross Farm Income*, 65; U.S. Census Bureau, *Fifteenth Census: 1930: Agriculture*, Part 2: 1089.

81. "Biggest Cotton Plantation," 131–132.

82. Robert W. Harrison, *Alluvial Empire* (Little Rock, Ark.: Pioneer Press, 1961), Vol. 1: 159–160; Arthur D. Frank, *The Development of the Federal Program of Flood Control on the Mississippi River* (New York, N.Y.: AMS Press, 1968), 213.

83. Harold Speakman, *Mostly Mississippi* (New York, N.Y.: Dodd, Mead and Co., 1927), 173, 241.

84. Maj. Gen. Edgar Jadwin, "The Plan for Flood Control of the Mississippi River and Its Alluvial Valley," *Annals of the American Academy of Political and Social Science*, 135(January 1928): 35–36; Harrison, *Levee Districts and Levee Building in Mississippi*, 235, 237; Frank, *Flood Control on the Mississippi River*, 237, 243, 246, 252.

85. "Biggest Cotton Plantation," 132; Strauss and Bean, *Gross Farm Income*, 65.

86. Moses S. Musoke, "Mechanizing Cotton Production in the American South: The Tractor, 1915–1960," *Explorations in Economic History*, 18(October 1981): 352–353.

87. "Mississippi Farm Empire: Ex-Field-hand Johnie [*sic*] Brown Owns Six-Mile Long Plantation Along Mississippi River Levee," *Ebony*, 6(February 1951): 33.

88. Vaiden and others, *Making Cotton Cheaper*, 8.

89. Vaiden and others, *Making Cotton Cheaper*, 8–15.

90. See Lewis E. Long and R. S. Kifer, *Farm Practices in South Central Mississippi with Suggested Changes*, Mississippi Agricultural Experiment Station, Bulletin No. 276 (Starkville, Miss.: Mississippi A & M College, 1929), 45; Musoke, "Mechanizing Cotton Production," 349.

91. William J. Carson, "Banking in the South," *Annals of the American Academy of Political and Social Science*, 153(January 1931): 213–214; Ben B. McNew, "Banking, 1890–1970," in McLemore, ed., *A History of Mississippi*, Vol. 2: 322.

92. McNair, *Labor Requirements of Arkansas Crops*, 9, 11.

93. Vaiden and others, *Making Cotton Cheaper*, 22.

94. See Clarence A. Wiley, "The Rust Mechanical Cotton Picker and the Probable Land-Use Adjustments," *Journal of Land and Public Utility Economics*, 15(May 1939): 155–157; Heywood Fleisig, "Mechanizing the Cotton Harvest in the Nineteenth-Century South," *Journal of Economic History*, 25(December 1965): 704–706; Hurt, "Agricultural Technology in the Twentieth Century," 30–38.

95. Alexander, *The Arkansas Plantation, 1920–1942*, 11–12, 16.

96. See Smith, *Soy Beans in Systems of Farming*, 4; Calhoun, "History of Concordia Parish," VIII, 605; Alvin A. Munn, "Production and Utilization of the Soybean in the United States," *Economic Geography*, 26(July 1950): 228–230; Harry D. Fornari, "The Big Change: Cotton to Soybeans," *Agricultural History*, 53(January 1979): 246, 248.

97. S. Earl Grigsby and Harold Hoffsommer, *Cotton Plantation Laborers: A Socio-Economic Study of Laborers on Cotton Plantations in Concordia Parish, Louisiana*, Louisiana Agricultural Experiment Station, Bulletin No. 328 (Baton Rouge, La.: Louisiana State University, 1941), 8–9.

98. See T. Lynn Smith, "The Redistribution of the Negro Population of the United States, 1910–1960," *Journal of Negro History*, 51(July 1966): 159; Robert Higgs, "The Boll Weevil, the Cotton Economy, and Black Migration, 1910–1930," *Agricultural History*, 50(April 1976): 343–345, 349–350; Neil Fligstein, *Going North: Migration of Blacks and Whites from the South, 1900–1950* (New York, N.Y.: Academic Press, 1981), 15–16, 131–133; Kirby, *Rural Worlds Lost*, 312, 316; Stewart E. Tolnay and E. M. Beck, "Racial Violence in the American South, 1910 to 1930," *American Sociological Review*, 57(February 1992): 112–113.

99. Lyonel Florant, "Negro Internal Migration," *American Sociological Review*, 7(December 1942): 784–785; Smith, "Redistribution of the Negro Population," 168–169.

100. Cobb, *Most Southern Place on Earth*, 115.

101. David L. Cohn, *Where I Was Born and Raised* (Boston, Mass.: Houghton Mifflin Co., 1948), 72–73.

102. See John S. Otto, *The Southern Frontiers*, 1607–1860 (Westport, Conn.: Greenwood Press, 1989), 2–3.

103. Herbert S. Sternitzke, "Impact of Changing Land Use in Delta Hardwood Forests," *Journal of Forestry*, 74(January 1976): 25; Purificacion O. MacDonald and others, "Documentation, Chronology, and Future Projections of Bottomland Hardwood Habitat Loss in the Lower Mississippi Alluvial Plain" (Washington, D.C.: Division of Ecological Services, U.S. Department of the Interior, 1979), Vol. 1: iv–v.

104. Philip E. Jones and others, *New Settlement Problems in the Northeastern Louisiana Delta*, Louisiana Agricultural Experiment Station, Bulletin No. 335 (Baton Rouge, La.: Louisiana State University, 1942), 6–7.

105. White and others, *Rich Land—Poor People*, 1–3.

106. Jonathan Daniels, *A Southerner Discovers the South* (New York, N.Y.: MacMillan Co., 1938), 187.

107. Bratton, "Geography of the St. Francis Basin," 27 note 55.

108. The living conditions of tenant farmers in the Southern bottomlands are described in Dorothy Dickins, *A Nutrition Investigation of Negro Tenants in the Yazoo Mississippi Delta*, Mississippi Agricultural Experiment Station, Bulletin No. 254 (Starkville, Miss.: Mississippi A & M College, 1928), 9–12, 17, 26–32 and White and others, *Rich Land—Poor People*, 37–49.

CHAPTER 5

1. Raymond C. Smith, "Ol' Man River's New Homes," *Land Policy Review*, 4(February 1941): 3.

2. Frederick Strauss and Louis Bean, *Gross Farm Income and Indices of Farm Production and Prices in the United States*, 1869–1937, U.S. Department of Agriculture, Technical Bulletin No. 703 (Washington, D.C.: Government Printing Office, 1940), 65–66.

3. "The Year in Agriculture," in U.S. Department of Agriculture, *Yearbook of*

the U.S. Department of Agriculture: 1931 (Washington, D.C.: Government Printing Office, 1931), 1; *Relief Work in the Drought of 1930–31* (Washington, D.C.: American National Red Cross, 1931), 8–9.

4. "The Great Drought of 1930," *Literary Digest*, 106(August 16, 1930): 5; "The 'Self-Help' Drought-Relief Program," *Literary Digest*, 106(August 30, 1930): 9; "Drought: Field Reports from Five of the States Most Seriously Affected," *New Republic*, 66(February 25, 1931): 40; Paul O. Kellogg, "Drought and the Red Cross," *Survey*, 65(February 15, 1931): 535; "The Year in Agriculture," 3–5; David E. Hamilton, "Herbert Hoover and the Great Drought of 1930," *Journal of American History*, 68(March 1982): 854, 856.

5. "Arkansas' Fight for Life," *Literary Digest*, 108(February 28, 1931): 5; *Relief Work in the Drought*, 21, Appendix X; Roger Lambert, "Hoover and the Red Cross in the Arkansas Drought of 1930," *Arkansas Historical Quarterly*, 29(Spring 1970): 5, 7, 16.

6. Nan E. Woodruff, *As Rare As Rain: Federal Relief in the Great Southern Drought of 1930–31* (Urbana, Ill.: University of Illinois Press, 1985), 28–29.

7. See "Stricken Arkansas," *Outlook and Independent*, 157(January 31, 1931): 85; A. L. Schafer, "When Hunger Followed Drought," *Survey*, 65(March 1, 1931): 581; Woodruff, *As Rare As Rain*, 29, 102.

8. Charles M. Wilson, "Famine in Arkansas," *Outlook and Independent*, 157(April 29, 1931): 596; "Drought: Field Reports," 40–41; A. Robbins, "Hunger—1931," *Nation*, 132(February 11, 1931): 152; *Relief Work in the Drought*, 14.

9. "Stricken Arkansas," 85; Kellogg, "Drought and the Red Cross," 535–536; Woodruff, *As Rare As Rain*, 45–50.

10. *Relief Work in the Drought*, 37–38.

11. "Now the 'Farm Breadline,' " *Literary Digest*, 108(January 24, 1931): 9; Lement Harris, "An Arkansas Farmer Speaks," *New Republic*, 67(May 27, 1931): 40–41; *Relief Work in the Drought*, 39–40; Woodruff, *As Rare As Rain*, 56–58.

12. Harris, "An Arkansas Farmer Speaks," 41.

13. "The Row Over Feeding Our Hungry Farmers," *Literary Digest*, 108(January 31, 1931): 8–9; Woodruff, *As Rare As Rain*, 66.

14. *Relief Work in the Drought*, 21, 29–37.

15. *Relief Work in the Drought*, Appendix III, Appendix IX.

16. Jeannie Whayne, *A New Plantation South* (Charlottesville, Va.: University Press of Virginia, 1996), 152–154.

17. "Now the 'Farm Breadline,' " 9; "Arkansas' Fight for Life," 5; Wilson, "Famine in Arkansas," 596; Lambert, "Hoover and the Red Cross," 14.

18. *Relief Work in the Drought*, 47–49, 67–68, Appendix X; Schafer, "When Hunger Followed Drought," 628; "Good News from Arkansas," *Literary Digest*, 108(March 28, 1931): 12; Burton E. Vaughan, "Arkansas Makes a Brilliant Recovery," *American Review of Reviews*, 83(June 1931): 91.

19. "Drought: Field Reports," 41.

20. Wilson, "Famine in Arkansas," 596; Vaughan, "Brilliant Recovery," 90.

21. *Relief Work in the Drought*, 54, 56.

22. "Drought: Field Reports," 37.

23. *Relief Work in the Drought*, 55.

24. Woodruff, *As Rare As Rain*, 135–136.

25. Harris, "An Arkansas Farmer Speaks," 41.

26. Woodruff, *As Rare As Rain*, 92, 98 note 80, 135.

27. Vaughan, "Brilliant Recovery," 90.

28. Wilson, "Famine in Arkansas," 597.

29. "Drought: Field Reports," 39.

30. Thad Snow, *From Missouri* (Boston, Mass.: Houghton Mifflin Co., 1954), 160–171.

31. Snow, *From Missouri*, 172–173; *Relief Work in the Drought*, Appendix IX.

32. "Drought: Field Reports," 37–38; *Relief Work in the Drought*, 27, Appendix IX.

33. "Drought: Field Reports," 37.

34. *Relief Work in the Drought*, 27.

35. "Drought: Field Reports," 37; Woodruff, *As Rare As Rain*, 130–131.

36. *Relief Work in the Drought*, 14.

37. "Drought: Field Reports," 38.

38. "Drought: Field Reports," 40; Woodruff, *As Rare As Rain*, 130.

39. Renwick C. Kennedy, "The Cotton Kingdom: 1931: I: Six-Cent Cotton, a Southern Tragedy," *New Republic*, 69(December 16, 1931): 129.

40. Kennedy, "Cotton Kingdom," 129.

41. Kennedy, "Cotton Kingdom," 129.

42. J. Wayne Flynt, *Dixie's Forgotten People: The South's Poor Whites* (Bloomington, Ind.: Indiana University Press, 1979), 67.

43. See Whayne, *New Plantation South*, 141–142.

44. Donald C. Alexander, *The Arkansas Plantation*, 1920–1942 (New Haven, Conn.: Yale University Press, 1943), 29.

45. Strauss and Bean, *Gross Farm Income*, 65–66; Gail S. Murray, "Forty Years Ago: The Great Depression Comes to Arkansas," *Arkansas Historical Quarterly*, 29(Winter 1970): 305–307; Whayne, *New Plantation South*, 158.

46. See Alexander, *Arkansas Plantation*, 26–28; Gerald D. Nash, "Herbert Hoover and the Origins of the Reconstruction Finance Corporation," *Mississippi Valley Historical Review*, 46(December 1959): 466–467; Van L. Perkins, *Crisis in Agriculture: The Agricultural Adjustment Administration and the New Deal, 1933*, University of California Publications in History, Vol. 81 (Berkeley, Calif.: University of California Press, 1969), 25–26; Murray, "Forty Years Ago, 308–309.

47. Alexander, *Arkansas Plantation*, 32; Perkins, *Crisis in Agriculture*, 35.

48. Perkins, *Crisis in Agriculture*, 36–39, 43, 49, 54, 69, 74–75, 101–103.

49. David E. Conrad, *The Forgotten Farmers: The Story of Sharecroppers in the New Deal* (Urbana, Ill.: University of Illinois Press, 1965), 37–38, 45; Perkins, *Crisis in Agriculture*, 91–93; Jack T. Kirby, *Rural Worlds Lost* (Baton Rouge, La.: Louisiana State University Press, 1987), 60.

50. Perkins, *Crisis in Agriculture*, 103–104.

51. Snow, *From Missouri*, 179–180.

52. Fred C. Frey and T. Lynn Smith, "The Influence of the AAA Cotton Program Upon the Tenant, Cropper, and Laborer," *Rural Sociology*, 1(December 1936): 488; Perkins, *Crisis in Agriculture*, 113–114; James C. Cobb, *The Most Southern Place on Earth* (New York, N.Y.: Oxford University Press, 1992), 186.

53. Conrad, *Forgotten Farmers*, 52; Perkins, *Crisis in Agriculture*, 114–115; Donald H. Grubbs, *Cry from the Cotton: The Southern Tenant Farmers' Union and the New Deal* (Chapel Hill, N.C.: University of North Carolina Press, 1971), 21.

54. Harold Hoffsommer, "The AAA and the Cropper," *Social Forces*, 13(May 1935): 497.

55. William R. Amberson, "The New Deal for Share-Croppers," *Nation*, 140(February 13, 1935): 185; Hoffsommer, "The AAA and the Cropper," 498; Conrad, *Forgotten Farmers*, 55–56; Cobb, *Most Southern Place on Earth*, 187–188.

56. Amberson, "New Deal for Share-Croppers," 186; H. L. Mitchell and J. R. Butler, "The Cropper Learns His Fate," *Nation*, 141(September 18, 1935): 328; Conrad, *Forgotten Farmers*, 76–77.

57. Amberson, "New Deal for Share-Croppers," 185; Conrad, *Forgotten Farmers*, 78–79; Grubbs, *Cry from the Cotton*, 21–22.

58. Conrad, *Forgotten Farmers*, 74–75.

59. Amberson, "New Deal for Share-Croppers," 187; Conrad, *Forgotten Farmers*, 80, 84.

60. See Mitchell and Butler, "The Cropper Learns His Fate," 328; Conrad, *Forgotten Farmers*, 84–87; Jerold S. Auerbach, "Southern Tenant Farmers: Socialist Critics of the New Deal," *Labor History*, 7(Winter 1966): 3–9; Mark D. Naison, "Black Agrarian Radicalism in the Great Depression: The Threads of a Lost Tradition," *Journal of Ethnic Studies*, 4(Fall 1973): 56; H. L. Mitchell, *Mean Things Happening in This Land: The Life and Times of H. L. Mitchell, Co-Founder of the Southern Tenant Farmers' Union* (Montclair, N.J.: Allanheld, Osman and Co., 1979), 46–52; Whayne, *New Plantation South*, 184–187, 197–198.

61. Lucien Koch, "The War in Arkansas," *New Republic*, 82(March 27, 1935): 184.

62. John Herling, "Field Notes from Arkansas," *Nation*, 140(April 10, 1935): 419–420; Koch, "War in Arkansas," 183–184; H. L. Mitchell and Howard Kester, "Correspondence: Share-Cropper Misery and Hope," *Nation*, 142(February 12, 1936): 185; Conrad, *Forgotten Farmers*, 95–97, 158–163; Mitchell, *Mean Things Happening in This Land*, 72–81.

63. "Terror in Arkansas," *Nation*, 140(February 13, 1935): 174; John Herling, "The Sharecroppers' Fight for Life," *New Republic*, 85(January 29, 1936): 336; Conrad, *Forgotten Farmers*, 166, 172–173; Mitchell, *Mean Things Happening in This Land*, 81–86.

64. Strauss and Bean, *Gross Farm Income*, 65.

65. Alexander, *Arkansas Plantation*, 42–43; Conrad, *Forgotten Farmers*, 203.

66. Strauss and Bean, *Gross Farm Income*, 65; Alexander, *Arkansas Plantation*, 44–46.

67. See Alexander, *Arkansas Plantation*, 46–50; James H. Street, *The New Revolution in the Cotton Economy: Mechanization and Its Consequences* (Chapel Hill, N.C.: University of North Carolina Press, 1957), 45; Warren C. Whatley, "Labor for the Picking: The New Deal in the South," *Journal of Economic History*, 43(December 1983): Table 6.

68. Charles S. Johnson and others, *The Collapse of Cotton Tenancy: Summary of Field Studies and Statistical Surveys*, 1933–1935 (Chapel Hill, N.C.: University of North Carolina Press, 1935), 48–52; Kirby, *Rural Worlds Lost*, 63; Cobb, *Most Southern Place on Earth*, 192, 195–196.

69. Snow, *From Missouri*, 197.

70. "Biggest Cotton Plantation," *Fortune*, 15(March 1937): 132; Strauss and Bean, *Gross Farm Income*, 65; Lawrence J. Nelson, "Oscar Johnston, the New Deal,

and the Cotton Subsidy Payments Controversy, 1936–1937," *Journal of Southern History*, 40(August 1974): 405–407.

71. Lawrence J. Nelson, "Welfare Capitalism on a Mississippi Plantation in the Great Depression," *Journal of Southern History*, 50(May 1984): 226–230.

72. "Biggest Cotton Plantation," 158.

73. Nelson, "Welfare Capitalism on a Mississippi Plantation," 232–233, 244–247.

74. Jonathan Daniels, *A Southerner Discovers the South* (New York, N.Y.: MacMillan Co., 1938), 190.

75. "Biggest Cotton Plantation," 128–129, 156, 160.

76. Oscar Johnston, "Industrial Plan of the Delta and PineLand Company of Mississippi," *Journal of Social Hygiene*, 26(February 1940): 73–77.

77. Nelson, "Welfare Capitalism," 244–245.

78. Daniels, *A Southerner Discovers the South*, 192.

79. Daniels, *A Southerner Discovers the South*, 185.

80. H. W. Blalock, *Plantation Operations of Landlords and Tenants in Arkansas*, Arkansas Agricultural Experiment Station, Bulletin No. 339 (Fayetteville, Ark.: University of Arkansas, 1937), 4, 13–15.

81. Alexander, *Arkansas Plantation*, 89–91.

82. Blalock, *Plantation Operations*, 10–12.

83. Alexander, *Arkansas Plantation*, 92–94.

84. Snow, *From Missouri*, 197–198.

85. "Mississippi Farm Empire: Ex-Field-hand Johnie [*sic*] Brown Owns Six-Mile Long Plantation Along Mississippi River Levee," *Ebony*, 6(February 1951): 34.

86. Snow, *From Missouri*, 198.

87. "Some Facts That the Statistics Show," *Implement and Tractor*, 54(July 8, 1939): 42b–42c; Cobb, *Most Southern Place on Earth*, 189.

88. Lewis E. Long, *Farm Power in the Yazoo-Mississippi Delta*, Mississippi Agricultural Experiment Station, Bulletin No. 295 (Starkville, Miss.: Mississippi A & M College, 1931), 4–5, 7–9, 17–18, 27.

89. E. L. Langsdorf and B. H. Thibodeaux, *Plantation Organization and Operation in the Yazoo-Mississippi Delta Area*, U.S. Department of Agriculture, Technical Bulletin No. 682 (Washington, D.C.: Government Printing Office, 1939), 58–59.

90. Long, *Farm Power in the Yazoo–Mississippi Delta*, 28–29.

91. Langsdorf and Thibodeaux, *Plantation Organization*, 61–69; Alexander, *Arkansas Plantation*, 100–102, 108.

92. John W. White, *Combinations of Enterprises on Plantations in the Lower Arkansas Delta*, Arkansas Agricultural Experiment Station, Bulletin No. 449 (Fayetteville, Ark.: University of Arkansas, 1944), 13–16.

93. J. G. McNeely and Glen T. Barton, *Land Tenure in Arkansas: II: Change in Labor Organization on Cotton Farms*, Arkansas Agricultural Experiment Station, Bulletin No. 397 (Fayetteville, Ark.: University of Arkansas, 1940), 8–10, 17, 25–26.

94. Amberson, "The New Deal for Share-Croppers," 186.

95. Street, *New Revolution in the Cotton Economy*, 218–220.

96. "Wage Workers and Sharecroppers on Mississippi Plantations," *Monthly La-*

bor Review, 49(November 1939): 1105–1106; Alexander, *Arkansas Plantation*, 101, 108–109.

97. Kirby, *Rural Worlds Lost*, 67, 72.

98. Frey and Smith, "The Influence of the AAA Cotton Program," 503.

99. Street, *New Revolution in the Cotton Economy*, 219.

100. See Wilson Gee, "Acreage Reduction and the Displacement of Farm Labor," *Journal of Farm Economics*, 17(August 1935): 525; Johnson and others, *Collapse of Cotton Tenancy*, 55, 58, 61; Grubbs, *Cry from the Cotton*, 25; Donald Holley, *Uncle Sam's Farmers: The New Deal Communities in the Lower Mississippi Valley* (Urbana, Ill.: University of Illinois Press, 1975), 83.

101. Lawrence Westbrook, "The Program of Rural Rehabilitation of the FERA," *Journal of Farm Economics*, 17(February 1935): 89–93; Dan W. Pittman, "The Founding of Dyess Colony," *Arkansas Historical Quarterly*, 29(Winter 1970): 313–314; Holley, *Uncle Sam's Farmers*, 25–26.

102. Daniels, *A Southerner Discovers the South*, 143–144; Mabel F. Edrington, *History of Mississippi County, Arkansas* (Ocala, Fla.: Ocala Star-Banner, 1962), 379–381; Pittman, "Founding of Dyess Colony," 315–319; Holley, *Uncle Sam's Farmers*, 28, 34–45.

103. Christopher S. Wren, "The Restless Ballad of Johnny Cash," *Look*, 33(April 29, 1969): 71–72.

104. Will W. Alexander, "Rural Resettlement," *Southern Review*, 1(Annual 1935–35): 529–530; Rexford G. Tugwell, "The Resettlement Idea," *Agricultural History*, 33(October 1959): 159; Holley, *Uncle Sam's Farmers*, 46–47, 66–81, 223–226, 284–285.

105. Mitchell and Kester, "Correspondence: Share-Cropper Misery and Hope," 185; "Woman Flogged: Charge in Arkansas Share-Croppers' Strike Causes Flare-up," *Literary Digest*, 121(June 27, 1936): 29–30; Alva W. Taylor, "Tenancy Report Scores System," *Christian Century*, 52(December 23, 1936): 1730–1731; Holley, *Uncle Sam's Farmers*, 89–96; Mitchell, *Mean Things Happening in This Land*, 87–88, 129–130.

106. C. T. Carpenter, "King Cotton's Slaves: The Fate of the Share-Croppers Becomes a National Issue," *Scribner's Magazine*, 98(October 1935): 194–199; Holley, *Uncle Sam's Farmers*, 96–103; Mitchell, *Mean Things Happening in This Land*, 100.

107. T. Roy Reid, "Public Assistance to Low-Income Farmers of the South," *Journal of Farm Economics*, 21(February 1939): 188–193; Tugwell, "Resettlement Idea," 160–161; Grubbs, *Cry from the Cotton*, 147–157; Holley, *Uncle Sam's Farmers*, 103–104, 277.

108. Donald Holley, "The Negro in the New Deal Resettlement Program," *Agricultural History*, 45(July 1971): 183–184; Holley, *Uncle Sam's Farmers*, 180–183, 284–285.

109. Jonathan Mitchell, "Cabins in the Cotton," *New Republic*, 92(September 22, 1937): 175.

110. Sherwood Eddy, "The Delta Cooperative's First Year," *Christian Century*, 54(February 3, 1937): 139–140; Mitchell, "Cabins in the Cotton," 175–176; Daniels, *A Southerner Discovers the South*, 148–155; Mitchell, *Mean Things Happening in This Land*, 131–133.

111. Mitchell, "Cabins in the Cotton," 175–176.

112. Reinhold Niebuhr, "Meditations from Mississippi," *Christian Century*, 54(February 10, 1937): 183–184.

113. See M. S. Venkataramani, "Norman Thomas, Arkansas Sharecroppers, and Roosevelt Agricultural Policies, 1933–1937," *Mississippi Valley Historical Review*, 47(September 1960): 244–245; Holley, "The Negro in the New Deal Resettlement Program," 192; Flynt, *Dixie's Forgotten People*, 84.

114. Whatley, "Labor for the Picking," 923–924.

115. Charles S. Hoffman and Virgil L. Bankson, "Crisis in Missouri's Boot Heel," *Land Policy Review*, 3(January–February 1940): 1–2; Oran Stephens, "Revolt on the Delta: What Happened to the Sharecroppers' Union," *Harper's Magazine*, 183(November 1941): 658–664.

116. Louis Cantor, "A Prologue to the Protest Movement: The Missouri Sharecropper Roadside Demonstration of 1939," *Journal of American History*, 55(March 1969): 810–811; Grubbs, *Cry from the Cotton*, 163–180; Mitchell, *Mean Things Happening in This Land*, 165–170.

117. Ben M. Ridpath, "The Case of the Missouri Sharecroppers," *Christian Century*, 56(February 1, 1939): 147; Cedric Belfrage, "Cotton-Patch Moses," *Harper's Magazine*, 197(November 1948): 98–99; Snow, *From Missouri*, 231–232, 235; Cantor, "A Prologue to the Protest Movement," 807–809.

The history of the LaForge resettlement project is discussed in Jeff Hearne, "The Beginning of LaForge: An Experiment in Rural Homesteading," *Missouri Historical Review*, 88(April 1994): 301–316.

118. See Ridpath, "The Case of the Missouri Sharecroppers," 146; Belfrage, "Cotton-Patch Moses," 100; Snow, *From Missouri*, 240–281; Cantor, "Prologue to the Protest Movement," 812–819; Mitchell, *Mean Things Happening in This Land*, 171–180; Arvah E. Strickland, "The Plight of the People in the Sharecroppers' Demonstration in Southeast Missouri," *Missouri Historical Review*, 81(July 1987): 403–416.

119. Belfrage, "Cotton-Patch Moses," 101–102; Snow, *From Missouri*, 281–283, 319–328; Cantor, "Prologue to the Protest Movement," 820–822; Amber T. Clifford, "To the Disinherited Belongs the Future," *Missouri Historical Review*, 98(October 1993): 26–28.

120. Strauss and Bean, *Gross Farm Income*, 71.

121. Pete Daniel, *Breaking the Land* (Urbana, Ill.: University of Illinois Press, 1985), 58, 134–150.

122. Murray, "Forty Years Ago," 306–307; Pamela Webb, "By the Sweat of the Brow: The Back-to-the-Land Movement in Depression Arkansas," *Arkansas Historical Quarterly*, 42(Winter 1983): 333–337.

123. C. Ray Brassieur, "From Two Mules to Twelve Row Equipment: An Oral History Interview with Maxwell Williams, Bootheel Farm Manager," *Missouri Historical Review*, 91(October 1996): 53–54, 61–65.

124. Smith, "Ol' Man River's New Homes," 3–4, 6.

125. U.S. Bureau of Agricultural Economics, prep., *New Settlement in the Mississippi Delta*, U.S. Department of Agriculture, Miscellaneous Publication No. 442 (Washington, D.C.: Government Printing Office, 1941), 2.

126. Margaret J. Bolsterli, *Born in the Delta: Reflections on the Making of a Southern White Sensibility* (Knoxville, Tenn.: University of Tennessee Press, 1991), 32.

127. Rudolf Heberle, "The Social Structure of New-Ground Settlements in the Mississippi Delta," *Social Forces*, 20(December 1941): 233–237.

128. Horace G. Porter, "New Farms in the Mississippi Delta," *Louisiana Rural Economist*, 2(April 1940): 5–7; U.S. Bureau of Agricultural Economics, prep., *New Settlement in the Mississippi Delta*, 6–7, 10, 12–13; Homer L. Hitt, "A Comparative Analysis of the People on New Ground Farms, Plantations, and Old Family Farms in the Upper Mississippi Delta of Louisiana," *Rural Sociology*, 7(December 1942): 407–410; Philip E. Jones and others, *New Settlement Problems in the Northeastern Louisiana Delta*, Louisiana Agricultural Experiment Station, Bulletin No. 335 (Baton Rouge, La.: Louisiana State University, 1942), 5–11, 14–15, 20–24, 26–29, 38.

129. U.S. Bureau of Agricultural Economics, prep., *New Settlement in the Mississippi Delta*, 1.

130. Smith, "Ol' Man River's New Homes," 7–8; Philip E. Jones and others, "New Settlement in the Delta of the Lower Mississippi Valley," *Journal of Land and Public Utility Economics*, 17(November 1941): 468–470.

131. Nicholas Lemann, *The Promised Land : The Great Black Migration and How It Changed America* (New York, N.Y.: Alfred A. Knopf, 1991), 3–5; Cobb, *Most Southern Place on Earth*, 198–200, 204.

132. Oscar Johnston, "Will the Machine Ruin the South?," *Saturday Evening Post*, 219(May 31, 1947): 37, 94–95.

133. Johnston, "Will the Machine Ruin the South?," 36, 98.

134. The mechanization of the Southern cotton economy is discussed in detail in Gilbert C. Fite, "Recent Progress in the Mechanization of Cotton Production in the United States," *Agricultural History*, 24(January 1950): 25–28; Gilbert C. Fite, *Cotton Fields No More: Southern Agriculture, 1865–1980* (Lexington, Ky.: University Press of Kentucky, 1984), 180–195; and Daniel, *Breaking the Land*, 239–255.

Sources

About the South on Lines of the Illinois Central and Yazoo & Mississippi Valley Railroad. Chicago, Ill.: Illinois Central Railroad Co., 1904.

Alexander, Donald C. *The Arkansas Plantation, 1920–1942.* New Haven, Conn.: Yale University Press, 1943.

Alexander, Will W. "Rural Resettlement." *Southern Review,* 1(Annual 1935–36): 528–539.

Allen, T. Warren. "Highways and Highway Transportation." In U.S. Department of Agriculture, *Agriculture Yearbook: 1924,* 97–184. Washington, D.C.: Government Printing Office, 1925.

Amberson, William R. "The New Deal for Share-Croppers." *Nation,* 140(February 13, 1935): 185–187.

"Anarchy and the Red Cross." *Nation,* 132(February 11, 1931): 144.

Anderson, John Q., ed. *Brokenburn: The Journal of Kate Stone, 1861–1868.* Baton Rouge, La.: Louisiana State University Press, 1955.

Arkansas. St. Louis, Mo.: Woodward and Tiernan Printing Co., 1890.

Arkansas. Washington, D.C.: U.S. Railroad Administration, 1919.

Arkansas: Statistics and Information. Little Rock, Ark.: Missouri Pacific Railway Co., 1888.

Arkansas Digest: 1820 to Date. St. Paul, Minn.: West Publishing Co., 1970.

Arkansas Facts. Little Rock, Ark.: State Bureau of Mines, Manufactures, and Agriculture, 1924.

The Arkansas Handbook. Ft. Smith, Ark.: State Bureau of Mines, Manufactures, and Agriculture, 1924.

"Arkansas's Fight for Life." *Literary Digest,* 108(February 28, 1931): 5–6.

Auerbach, Jerold S. "Southern Tenant Farmers: Socialist Critics of the New Deal." *Labor History,* 7(Winter 1966): 3–18.

Ayers, Edward L. *Southern Crossing: A History of the American South, 1877–1906.* New York, N.Y.: Oxford University Press, 1995.

Baker, Oliver E. "Agricultural Regions of North America: Part I: The Basis of Classification." *Economic Geography*, 2(July 1926): 459–493.

Baker, Oliver E. "Agricultural Regions of North America: Part II: The South." *Economic Geography*, 3(January 1927): 50–86.

Baker, Oliver E. "Agricultural Regions of North America: Part IV: The Corn Belt." *Economic Geography*, 3(October 1927): 447–465.

Baker, Ray S. *Following the Color Line: An Account of Negro Citizenship in the American Democracy.* New York, N.Y.: Doubleday, Page and Co., 1908.

Barclay, James R., and Alexander Keith. *The Aberdeen-Angus Breed: A History.* Aberdeen, U.K.: Aberdeen Angus Cattle Society, 1958.

Barry, John M. *Rising Tide: The Great Mississippi Flood of 1927 and How It Changed America.* New York, N.Y.: Simon and Schuster, 1997.

Bedenbaugh, P. G. *Grazing and Feeding Trials with Corn and Soybeans for Pork Production.* Mississippi Agricultural Experiment Station, Bulletin No. 283. Starkville, Miss.: Mississippi A & M College, 1930.

Belfrage, Cedric. "Cotton-Patch Moses." *Harper's Magazine*, 197(November 1948): 94–102.

Bennett, Hugh H. *The Soils and Agriculture of the Southern States.* New York, N.Y.: MacMillan Co., 1921.

Bennett, R. L. *A Method of Breeding Early Cotton to Escape Boll-Weevil Damage.* U.S. Department of Agriculture, Farmers' Bulletin No. 314. Washington, D.C.: Government Printing Office, 1908.

"Biggest Cotton Plantation." *Fortune*, 15(March 1937): 125–132, 156–160.

Black, Robert C., III. *The Railroads of the Confederacy.* Chapel Hill, N.C.: University of North Carolina Press, 1952.

Blalock, H. W. *Plantation Operations of Landlords and Tenants in Arkansas.* Arkansas Agricultural Experiment Station, Bulletin No. 339. Fayetteville, Ark.: University of Arkansas, 1937.

Boeger, E. A., and E. A. Goldenweiser. *A Study of the Tenant Systems of Farming in the Yazoo–Mississippi Delta.* U.S. Department of Agriculture, Bulletin No. 337. Washington, D.C.: Government Printing Office, 1916.

Boles, John B. *Black Southerners, 1619–1869.* Lexington, Ky.: University Press of Kentucky, 1984.

Bolsterli, Margaret J. *Born in the Delta: Recollections on the Making of a Southern White Sensibility.* Knoxville, Tenn.: University of Tennessee Press, 1991.

"Boost Missouri" and "The Call of the Farm." Jefferson City, Mo.: Missouri Bureau of Labor Statistics, [1915].

Bowman, Robert. "Early History and Archaeology of Yazoo County." *Publications of the Mississippi Historical Society*, 8(Annual 1904): 427–441.

Boyd, George R. *Clearing Land of Brush and Stumps.* U.S. Department of Agriculture, Farmers' Bulletin No. 1526. Washington, D.C.: Government Printing Office, 1927.

Boyd, Mark F. "An Historical Sketch of the Prevalence of Malaria in North America." *American Journal of Tropical Medicine*, 21(March 1941): 223–244.

Brandfon, Robert L. *Cotton Kingdom of the New South: A History of the Yazoo Mis-*

sissippi Delta from the Reconstruction to the Twentieth Century. Cambridge, Mass.: Harvard University Press, 1967.

Brassieur, C. Ray. "From Two Mules to Twelve-Row Equipment: An Oral History Interview with Maxwell Williams, Bootheel Farm Manager." *Missouri Historical Review*, 91(October 1996): 52–85.

Bratton, Samuel T. "The Geography of the St. Francis Basin." *University of Missouri Studies*, 1(July 1926): 1–54.

Braun, E. Lucy. *Deciduous Forests of Eastern North America*. New York, N.Y.: Free Press, 1974.

Brough, Charles H. "History of Taxation in Mississippi." *Publications of the Mississippi Historical Society*, 2(Annual 1899): 113–124.

Brown, Robert M. "The Mississippi River as a Trade Route." *Bulletin of the American Geographical Society*, 38(June 1906): 349–354.

Brown, Walter L. "Notes on Early Rice Culture in Arkansas." *Arkansas Historical Quarterly*, 29(Spring 1970): 76–79.

Browne, W. A. "Grand Prairie: A Progressive Rice Region." *Journal of Geography*, 31(April 1932): 137–147.

Calhoun, R. D. "A History of Concordia Parish, Louisiana," II, III, IV, VI, VII and VIII, *Louisiana Historical Quarterly*, 15(April 1932): 214–233; 15(July 1932): 428–452; 15(October 1932): 618–645; 16(April 1933): 309–329; 16(July 1933): 454–478; 16(October 1933): 598–607.

Cantor, Louis. "A Prologue to the Protest Movement: The Missouri Sharecropper Roadside Demonstration of 1939." *Journal of American History*, 55(March 1969): 804–822.

Carpenter, C. T. "King Cotton's Slaves: The Fate of the Share-Cropper Becomes a National Issue." *Scribner's Magazine*, 98(October 1935): 193–199.

Carrier, Lyman. *Lespedeza as a Forage Crop*. U.S. Department of Agriculture, Farmer's Bulletin No. 1143. Washington, D.C.: Government Printing Office, 1920.

Carrier, Lyman, and F. G. Ashbrook. *Hog Pastures for the Southern States*. U.S. Department of Agriculture, Farmers' Bulletin No. 951. Washington, D.C.: Government Printing Co., 1918.

Carson, William J. "Banking in the South." *Annals of the American Academy of Political and Social Science*, 153(January 1931): 210–223.

Cates, H. R. *Farm Practice in the Cultivation of Corn*. U.S. Department of Agriculture, Bulletin No. 320. Washington, D.C.: Government Printing Office, 1916.

Cates, H. R. *Farm Practice in the Cultivation of Cotton*. U.S. Department of Agriculture, Department Bulletin No. 511. Washington, D.C.: Government Printing Office, 1917.

Chilton, Andrew G. *Through Arkansas on the Hog*. Chicago, Ill.: M. A. Donohue and Co., 1908.

Clark, Ira G. *Then Came the Railroads: The Century from Steam to Diesel in the Southwest*. Norman, Okl.: University of Oklahoma Press, 1958.

Clark, Thomas D. "Changes in Transportation." In *A History of Mississippi* Ed. by Richard A. McLemore, Vol. 2: 274–311. Jackson, Miss.: University Press of Mississippi, 1973.

Clements, William M., and Larry D. Ball. "This Was the Beginning of Clearing of

Land: The Development and Use of the East Arkansas Stump Saw." *Arkansas Historical Quarterly*, 45(Spring 1986): 41–52.

Clifford, Amber T. "To the Disinherited Belongs the Future." *Missouri Historical Review*, 88(October 1993): 22–28.

Coad, B. R. "Airplane Dusting of Cotton Fields Proves Effective, Economical." In U.S. Department of Agriculture, *Yearbook of Agriculture: 1928*, 117–120. Washington, D.C.: Government Printing Office, 1929.

Coad, B. R., and T. P. Cassidy. *Cotton Boll-Weevil Control by the Use of Poison*. U.S. Department of Agriculture, Department Bulletin No. 875. Washington, D.C.: Government Printing Office, 1920.

Cobb, James C. *The Most Southern Place on Earth: The Mississippi Delta and the Roots of Regional Identity*. New York, N.Y.: Oxford University Press, 1992.

Cohn, David L. *Where I Was Born and Raised*. Boston, Mass.: Houghton Mifflin Co., 1948.

Conrad, David E. *The Forgotten Farmers: The Story of Sharecroppers in the New Deal*. Urbana, Ill.: University of Illinois Press, 1965.

Cook, O. F. *Cotton Improvement Under Weevil Conditions*. U.S. Department of Agriculture, Farmers' Bulletin No. 501. Washington, D.C.: Government Printing Office, 1912.

Creed, David A. "An Economic and Social History of Madison Parish, Louisiana." M. A. thesis, Northeast Louisiana State College, Monroe, 1969.

Currie, James T. *Enclave: Vicksburg and Her Plantations, 1863–1870*. Jackson, Miss.: University Press of Mississippi, 1980.

Daniel, Pete. *The Shadow of Slavery: Peonage in the South, 1901–1969*. Urbana, Ill.: University of Illinois Press, 1972.

Daniel, Pete. *Deep 'n As It Come: The 1927 Mississippi River Flood*. New York, N.Y.: Oxford University Press, 1977.

Daniel, Pete. *Breaking the Land: The Transformation of Cotton, Tobacco, and Rice Cultures Since 1880*. Urbana, Ill.: University of Illinois Press, 1985.

Daniels, Jonathan. *A Southerner Discovers the South*. New York, N.Y.: MacMillan Co., 1938.

Darling, J. N. "The Yazoo Delta of Mississippi and Location and Construction of Its Railroads." *Journal of the Western Society of Engineers*, 7(December 1902): 560–573.

Davis, Burke. *The Southern Railway: Road of the Innovators*. Chapel Hill, N.C.: University of North Carolina Press, 1985.

Davis, P. O. "The Negro Exodus and Southern Agriculture." *American Review of Reviews*, 68(October 1923): 401–407.

Davis, Robert M. *The Southern Planter, the Factor, and the Banker*. New Orleans, La.: R. M. Davis, 1871.

Davis, Ronald L. F. *Good and Faithful Labor: From Slavery to Sharecropping in the Natchez District, 1860–1890*. Westport, Conn.: Greenwood Press, 1982.

"The Delta of Northwestern Mississippi." *DeBow's Review*, o.s. 13(November 1852): 475–477.

Dethloff, Henry C. "Rice Revolution in the Southwest, 1880–1910." *Arkansas Historical Quarterly*, 29(Spring 1970): 66–75.

Dethloff, Henry C. *A History of the American Rice Industry, 1685–1985*. College Station, Tex.: Texas A & M University Press, 1988.

Dew, Lee A. "The J.L.C. and E.R.R. [Jonesboro, Lake City & Eastern Railroad] and the Opening of the 'Sunk Lands' of Northeast Arkansas." *Arkansas Historical Quarterly*, 27(Spring 1968): 22–39.

Dew, Lee A., and Louis Koeppe. "Narrow-Gauge Railroads in Arkansas." *Arkansas Historical Quarterly*, 31(Autumn 1972): 276–293.

Dial, Marshall. *The Bootheel Swamp Struggle.* Lilbourn, Mo.: Lloyd Publications, 1961.

Dickins, Dorothy. *A Nutrition Investigation of Negro Tenants in the Yazoo Mississippi Delta.* Mississippi Agricultural Experiment Station, Bulletin No. 254. Starkville, Miss.: Mississippi A & M College, 1928.

Doherty, William T., Jr. *Louis Houck, Missouri Historian and Entrepreneur.* University of Missouri Studies, Vol. 33. Columbia, Mo.: University of Missouri Press, 1960.

Donaldson, Thomas C. *The Public Domain: Its History with Statistics.* Washington, D.C.: Government Printing Office, 1884.

Dorris, Fern E. "The Yazoo Basin in Mississippi." *Journal of Geography*, 28(February 1929): 72–79.

Dorset, M. *Hog Cholera.* U.S. Department of Agriculture, Farmers' Bulletin No. 379. Washington, D.C.: Government Printing Office, 1909.

Doty, S. W. *Marketing Live Stock in the South: Suggestions for Improvement.* U.S. Department of Agriculture, Farmers' Bulletin No. 809. Washington, D.C.: Government Printing Office, 1917.

Douglass, Robert S. *History of Southeast Missouri.* Chicago, Ill.: Lewis Publishing Co., 1912.

"Down the Mississippi." *Harper's New Monthly Magazine*, 41(November 1870): 835–845.

Downs, Eldon W., and George F. Lemmer. "Origins of Aerial Crop Dusting." *Agricultural History*, 39(July 1965): 123–135.

"Drought: Field Reports from Five of the States Most Seriously Affected." *New Republic*, 66(February 25, 1931): 37–41.

Economic Survey of Arkansas. St. Louis, Mo.: Southwestern Bell Telephone Co., 1925.

Economic Survey of Missouri. St. Louis, Mo.: Southwestern Bell Telephone Co., 1927.

Eddy, Sherwood. "The Delta Cooperative's First Year." *Christian Century*, 54(February 3, 1937): 139–140.

Edrington, Mabel F. *History of Mississippi County, Arkansas.* Ocala, Fla.: Ocala Star-Banner, 1962.

Eldridge, Maurice O. *Good Roads for Farmers.* U.S. Department of Agriculture, Farmers' Bulletin No. 95. Washington, D.C.: Government Printing Office, 1899.

Ellenberger, W. P., and Robert M. Chapin. *Cattle-Fever Ticks and Methods of Eradication.* U.S. Department of Agriculture, Farmers' Bulletin No. 1057. Washington, D.C.: Government Printing Office, 1919.

Experiment Station Work, XLII. U.S. Department of Agriculture, Farmers' Bulletin No. 305. Washington, D.C.: Government Printing Office, 1907.

Farley, F. W. "Growth of the Beef-Cattle Industry in the South." In U.S. Department of Agriculture, *Yearbook of the Department of Agriculture: 1917*, 327–340. Washington, D.C.: Government Printing Office, 1918.

Fauntleroy, Thomas. "Lord, Plant My Feet on Higher Ground: The Mississippi at Its Worst." *Independent*, 118(June 4, 1927): 580–586.

Faust, Ernest C. "The History of Malaria in the United States." *American Scientist*, 39(January 1951): 121–130.

Fite, Gilbert C. "Recent Progress in the Mechanization of Cotton Production in the United States." *Agricultural History*, 24(January 1950): 19–28.

Fite, Gilbert C. *Cotton Fields No More: Southern Agriculture, 1865–1980*. Lexington, Ky.: University Press of Kentucky, 1984.

Fleisig, Heywood. "Mechanizing the Cotton Harvest in the Nineteenth-Century South." *Journal of Economic History*, 25(December 1965): 704–706.

Fligstein, Neil. *Going North: Migration of Blacks and Whites from the South, 1900–1950*. New York, N.Y.: Academic Press, 1981.

Florant, Lyonel C. "Negro Internal Migration." *American Sociological Review*, 7(December 1942): 782–791.

Flynt, J. Wayne. *Dixie's Forgotten People: The South's Poor Whites*. Bloomington, Ind.: Indiana University Press, 1979.

Folsom, J. W. *Insect Enemies of the Cotton Plant*. U.S. Department of Agriculture, Farmers' Bulletin No. 1688. Washington, D.C.: Government Printing Office, 1932.

Fornari, Harry D. "The Big Change: Cotton to Soybeans." *Agricultural History*, 53(January 1979): 245–253.

Forshey, Col. Caleb G. *The Delta of the Mississippi*. Cambridge, Mass.: John Wilson and Son, 1873.

Fortune, T. Thomas. *Black and White: Land, Labor, and Politics in the South*. New York, N.Y.: Fords, Howard and Hulbert, 1884.

Foti, Thomas. "The River's Gifts and Curses." In *The Arkansas Delta: Land of Paradox*. Ed. by Jeannie Whayne and Willard Gatewood, 30–57. Fayetteville, Ark.: University of Arkansas Press, 1993.

"Four Stages of the Flood." *Survey*, 58(July 1, 1927): 357.

Frank, Arthur D. *The Development of the Federal Program of Flood Control on the Mississippi River*. New York, N.Y.: AMS Press, 1968.

Frey, Fred C., and T. Lynn Smith. "The Influence of the AAA Cotton Program Upon the Tenant, Cropper, and Laborer." *Rural Sociology*, 1(December 1936): 483–505.

Gates, Paul W. "The Railroads of Missouri, 1850–1870." *Missouri Historical Review*, 26(January 1932): 126–141.

Gatewood, Willard B., Jr. "Sunnyside: The Evolution of an Arkansas Plantation, 1840–1945." *Arkansas Historical Quarterly*, 50(Spring 1991): 5–29.

Gee, Wilson. "Acreage Reduction and the Displacement of Farm Labor." *Journal of Farm Economics*, 17(August 1935): 522–528.

Gerteis, Louis S. *From Contraband to Freedman*. Westport, Conn.: Greenwood Press, 1973.

Gibbens, Byrd. "Strangers in the Arkansas Delta: Ethnic Groups and Nationalities." In *The Arkansas Delta: Land of Paradox* Ed. by Jeannie Whayne and Willard Gatewood, 150–180. Fayetteville, Ark.: University of Arkansas Press, 1993.

Giles, William L. "Agricultural Revolution, 1890–1970." In *A History of Mississippi* Ed. by Richard A. McLemore, Vol. 2: 177–211. Jackson, Miss.: University Press of Mississippi, 1973.

"Good News from Arkansas." *Literary Digest*, 108(March 28, 1931): 12.
Gordon, Fon Louise. "From Slavery to Uncertain Freedom: Blacks in the Delta." In *The Arkansas Delta: Land of Paradox* Ed. by Jeannie Whayne and Willard Gatewood, 98–121. Fayetteville, Ark.: University of Arkansas Press, 1993.
Gray, Dan T. *Feeding Hogs in the South*. U.S. Department of Agriculture, Farmers' Bulletin No. 411. Washington, D.C.: Government Printing Office, 1910.
Gray, William F., ed. *Imperial Bolivar*. Cleveland, Miss.: Bolivar Commercial, 1923.
"The Great Drought of 1930." *Literary Digest*, 106(August 16, 1930): 5–6.
Grigsby, S. Earl, and Harold Hoffsommer. *Cotton Plantation Laborers: A Socio-Economic Study of Laborers on the Cotton Plantations in Concordia Parish, Louisiana*. Louisiana Agricultural Experiment Station, Bulletin No. 328. Baton Rouge, La.: Louisiana State University, 1941.
Grimmett, Harold K. "Early Plant and Animal Communities of the Arkansas Delta." *Arkansas Historical Quarterly*, 48(Summer 1989): 101–107.
Grubbs, Donald H. *Cry From the Cotton: The Southern Tenant Farmers' Union and the New Deal*. Chapel Hill, N.C.: University of North Carolina Press, 1971.
Hair, William I. *Bourbonism and Agrarian Protest: Louisiana Politics, 1877–1900*. Baton Rouge, La.: Louisiana State University Press, 1969.
Hall, Edward H. *Appleton's Hand-Book of American Travel: A Southern Tour*. New York, N.Y.: D. Appleton and Co., 1872.
Halsell, Willie D. "Migration Into, and Settlement of, Leflore County, 1833–1876." *Journal of Mississippi History*, 9(October 1947): 219–237.
Halsell, Willie D. "Migration Into, and Settlement of, Leflore County in the Later Periods, 1876–1920." *Journal of Mississippi History*, 10(July 1948): 240–260.
Hamilton, David E. "Herbert Hoover and the Great Drought of 1930." *Journal of American History*, 68(March 1982): 850–875.
Handley, Lawrence R. "Settlement Across Northern Arkansas as Influenced by the Missouri & North Arkansas Railroad." *Arkansas Historical Quarterly*, 33(Winter 1974): 273–292.
Harris, Lement. "An Arkansas Farmer Speaks." *New Republic*, 67(May 27, 1931): 40–41.
Harris, Wm. H. *Louisiana Products, Resources, and Attractions*. New Orleans, La.: New Orleans Democrat, 1881.
Harrison, Robert W. "Flood Control in the Yazoo-Mississippi Delta." *Southern Economic Journal*, 17(October 1950): 148–158.
Harrison, Robert W. "Levee Building in Mississippi Before the Civil War." *Journal of Mississippi History*, 12(April 1950): 63–97.
Harrison, Robert W. "The Formative Years of the Yazoo-Mississippi Delta Levee District." *Journal of Mississippi History*, 13(October 1951): 236–248.
Harrison, Robert W. *Levee Districts and Levee Building in Mississippi*. Stoneville, Miss.: Delta Council and others, 1951.
Harrison, Robert W. "Clearing Land in the Mississippi Alluvial Valley." *Arkansas Historical Quarterly*, 13(Winter 1954): 352–371.
Harrison, Robert W. *Alluvial Empire*. Little Rock, Ark.: Pioneer Press, 1961.
Harrison, Robert W., and Walter M. Kollmorgen. "Land Reclamation in Arkansas Under the Swamp Land Grant of 1850." *Arkansas Historical Quarterly*, 6(Winter 1947): 369–418.
Harrison, Robert W., and Walter M. Kollmorgen. *Drainage Reclamations in the*

Bartholomew-Boeuf-Tensas Basin of Arkansas and Louisiana. Arkansas Agricultural Experiment Station, Bulletin No. 476. Fayetteville, Ark.: University of Arkansas, 1948.

Harrison, Robert W., and Walter M. Kollmorgen. "Socio-Economic History of Cypress Creek Drainage District and Related Districts of Southeast Arkansas." *Arkansas Historical Quarterly*, 7(Spring 1948): 20–56.

Hart, Albert B. *The Southern South.* New York, N.Y.: D. Appleton and Co., 1910.

Hart, John F. "The Role of the Plantation in Southern Agriculture." In *Proceedings of the Tall Timbers Fire Ecology Conference, No. 16.* Tallahassee, Fla.: Tall Timbers Research Station, 1982.

Haskell, E. S. *Systems of Hog Farming in the Southeastern States.* U.S. Department of Agriculture, Farmers' Bulletin No. 985. Washington, D.C.: Government Printing Office, 1918.

Hataway, Marsha P. "The Development of the Mississippi State Highway System, 1916–1932," *Journal of Mississippi History*, 28(November 1966): 286–303.

Haygood, Tamara M. "Cow, Ticks, and Disease: A Medical Interpretation of the Southern Cattle Industry." *Journal of Southern History*, 52(November 1986): 551–564.

Hayter, Earl W. "Barbed Wire Fencing—A Prairie Invention." *Agricultural History*, 13(October 1939): 189–207.

Hearne, Jeff. "The Beginning of LaForge: An Experiment in Rural Homesteading." *Missouri Historical Review*, 88(April 1994): 301–316.

Helms, Douglas. "Technological Methods for Boll Weevil Control." *Agricultural History*, 53(January 1979): 286–299.

Heberle, Rudolf. "The Social Structure of New-Ground Settlements in the Mississippi Delta." *Social Forces*, 20(December 1941): 232–237.

Henry, James P. *Resources of the State of Arkansas* Little Rock, Ark.: Price and McClure, 1873.

Herling, John. "Field Notes from Arkansas." *Nation*, 140(April 10, 1935): 419–420.

Herling, John. "The Sharecroppers' Fight for Life." *New Republic*, 85(January 29, 1936): 336.

Hickman, Nollie W. "The Tensas-Bayou Macon." In *The Rivers and Bayous of Louisiana* Ed. by Edwin A. Davis, 14–20. Baton Rouge, La.: Louisiana Education Research Association, 1968.

Hickman, Nollie W. "Mississippi Forests." In *A History of Mississippi* Ed. By Richard A. McLemore, Vol. 2: 212–232. Jackson, Miss.: University Press of Mississippi, 1973.

Higgs, Robert. "The Boll Weevil, the Cotton Economy, and Black Migration, 1910–1930." *Agricultural History*, 50(April 1976): 335–350.

Highsmith, William E. "Louisiana Landholding During War and Reconstruction." *Louisiana Historical Quarterly*, 38(January 1955): 39–54.

Hill, George W. "The Bureau of Animal Industry." *Chautauquan*, 14(February 1892): 549–553.

A History of the North-Western Editorial Excursion to Arkansas. Little Rock, Ark.: T. B. Mills and Co., 1876.

Hitt, Homer L. "A Comparative Analysis of the People on New Ground Farms,

Plantations, and Old Family Farms in the Upper Mississippi Delta of Louisiana." *Rural Sociology*, 7(December 1942): 404–414.

Hoffman, Charles S., and Virgil L. Bankson. "Crisis in Missouri's Boot Heel." *Land Policy Review*, 3(January–February 1940): 1–14.

Hoffman, Charles. "The Depression of the Nineties." *Journal of Economic History*, 16(June 1956): 137–164.

Hoffsommer, Harold. "The AAA and the Cropper." *Social Forces*, 13(May 1935): 494–501.

Holley, Donald. "The Negro in the New Deal Resettlement Program." *Agricultural History*, 45(July 1971): 179–193.

Holley, Donald. *Uncle Sam's Farmers: The New Deal Communities in the Lower Mississippi Valley*. Urbana, Ill.: University of Illinois Press, 1975.

Holley, Donald. "The Plantation Heritage: Agriculture in the Arkansas Delta." In *The Arkansas Delta: Land of Paradox* Ed. by Jeannie Whayne and Willard Gatewood, 238–271. Fayetteville, Ark.: University of Arkansas Press, 1993.

Holmes, George K. *Cotton Crop of the United States, 1790–1911*. U.S. Department of Agriculture, Bureau of Statistics, Circular No. 32. Washington, D.C.: Government Printing Office, 1912.

Howard, L. O. *Insects Affecting the Cotton Plant*. U.S. Department of Agriculture, Farmers' Bulletin No. 47. Washington, D.C.: Government Printing Office, 1897.

Hudson, John C. "The Yazoo–Mississippi Delta as Plantation Country." In *Proceedings of the Tall Timbers Fire Ecology Conference, No. 16*. Tallahassee, Fla.: Tall Timbers Research Station, 1982.

Huff, Leo E. "The Memphis and Little Rock Railroad During the Civil War." *Arkansas Historical Quarterly*, 23(Autumn 1964): 260–270.

Hughes, Louis. *Thirty Years a Slave: From Bondage to Freedom*. Milwaukee, Wis.: SouthSide Publishing Co., 1897.

Hull, Clifton E. *Shortline Railroads of Arkansas*. Norman, Okl.: University of Oklahoma Press, 1969.

Hunter, Louis C. *Steamboats on the Western Rivers*. New York, N.Y.: Octagon Books, 1969.

Hunter, W. D. "The Present Status of the Mexican Boll-Weevil in the United States." In U.S. Department of Agriculture, *Yearbook of the U.S. Department of Agriculture: 1901*, 369–380. Washington, D.C.: Government Printing Office, 1902.

Hunter, W. D. "The Status of the Mexican Cotton Boll-Weevil in the United States in 1903." In U.S. Department of Agriculture, *Yearbook of the U.S. Department of Agriculture: 1903*, 205–214. Washington, D.C.: Government Printing Office, 1904.

Hunter, W. D. *The Control of the Boll-Weevil*. U.S. Department of Agriculture, Farmers' Bulletin No. 216. Washington, D.C.: Government Printing Office, 1905.

Hunter, W. D. "Present Status of the Cotton Boll-Weevil in the U.S." In U.S. Department of Agriculture, *Yearbook of the U.S. Department of Agriculture: 1904*, 191–204. Washington, D.C.: Government Printing Office, 1905.

Hunter, W. D. "Some Recent Studies of the Mexican Cotton Boll-Weevil." In U.S.

Department of Agriculture, *Yearbook of the U.S. Department of Agriculture: 1906*, 313–324. Washington, D.C.: Government Printing Office, 1907.

Hunter, W. D. "Relation Between Rotation Systems and Insect Injury in the South." In U.S. Department Agriculture, *Yearbook of the U.S. Department of Agriculture: 1911*, 201–210. Washington, D.C.: Government Printing Office, 1912.

Hunter, W. D. *The Boll-Weevil Problem.* U.S. Department of Agriculture, Farmers' Bulletin No. 848. Washington, D.C.: Government Printing Office, 1917.

Hunter, W. D., and B. R. Coad. *The Boll-Weevil Problem.* U.S. Department of Agriculture, Farmers' Bulletin No. 1262. Washington, D.C.: Government Printing Office, 1922.

Hurt, A. B. *Mississippi.* U.S. Department of Agriculture, Miscellaneous Special Report No. 3. Washington, D.C.: Government Printing Office, 1883.

Hurt, R. D. "Agricultural Technology in the Twentieth Century." *Journal of the West*, 30(April 1991): 3–100.

Hutchins, Lewis E., comp. *Resources of Arkansas.* Little Rock, Ark.: State Bureau of Mines, Manufactures, and Agriculture, 1911.

"Immigration: Opportunities for Farmers in the Mississippi Valley." *Southern Farm Magazine*, 6(June 1898): 17–18.

Industrial Commission. *Report of the Industrial Commission on Agriculture and Agricultural Labor, Including Testimony with Review and Topical Digest Thereof.* Washington, D.C.: Government Printing Office, 1900–1902.

An Invitation to Immigrants: Louisiana. Baton Rouge, La.: Louisiana Bureau of Immigration, 1894.

Irwin, Will. "Can We Tame the Mississippi?" *World's Work*, 54(August 1927): 405–415.

Irwin, Will. "The Aftermath—Mud and Money." *Survey*, 58(July 1, 1927): 358.

Jackson, Nannie S. *Vinegar Pie and Chicken Bread: A Woman's Diary of Life in the Rural South.* Ed. by Margaret Bolsterli. Fayetteville, Ark.: University of Arkansas, 1982.

Jadwin, Maj. Gen. Edgar. "The Plan for Flood Control of the Mississippi River in Its Alluvial Valley." *Annals of the American Academy of Political and Social Science*, 135(January 1928): 35–44.

James, D. Clayton. "Mississippi Agriculture, 1861–1865." *Journal of Mississippi History*, 24(July 1962): 129–141.

Jaynes, Gerald D. *Branches Without Roots: Genesis of the Black Working Class in the American South, 1862–1882.* New York, N.Y.: Oxford University Press, 1986.

Jenks, George F., and Robert W. Harrison. *Methods and Costs of Clearing Land in Northeast Arkansas.* Arkansas Agricultural Experiment Station, Bulletin No. 495. Fayetteville, Ark.: University of Arkansas, 1950.

Johnson, Charles S., and others. *The Collapse of Cotton Tenancy: Summary of Field Studies and Statistical Surveys, 1933–1935.* Chapel Hill, N.C.: University of North Carolina Press, 1935.

Johnson, Clifton. *Highways and Byways of the Mississippi Valley.* New York, N.Y.: MacMillan Co., 1906.

Johnson, Elmer, and B. R. Coad. *Dusting Machinery for Cotton Boll-Weevil Control.* U.S. Department of Agriculture, Farmers' Bulletin No. 1098. Washington, D.C.: Government Printing Office, 1920.

Johnston, Oscar. "Industrial Plan of the Delta and PineLand Company of Mississippi." *Journal of Social Hygiene*, 26(February 1940): 73–77.

Johnston, Oscar. "Will the Machine Ruin the South?" *Saturday Evening Post*, 219(May 31, 1947): 36–37, 94–98.

Jones, Philip E., and others. "New Settlement in the Delta of the Lower Mississippi Valley." *Journal of Land and Public Utility Economics*, 17(November 1941): 465–476.

Jones, Philip E., and others. *New Settlement Problems in the Northeastern Louisiana Delta*. Louisiana Agricultural Experiment Station, Bulletin No. 335. Baton Rouge, La.: Louisiana State University, 1942.

Keating, Bern. *A History of Washington County, Mississippi*. Greenville, Miss.: Greenville Junior Auxiliary, 1976.

Kelley, Arthell. "Levee Building and the Settlement of the Yazoo Basin." *Southern Quarterly*, 1(July 1963): 285–308.

Kelley, Arthell. "Sequent Occupancy of the Yazoo Basin, Mississippi, 1830 to 1976." *Mississippi Geographer*, 6(Spring 1978): 3–17.

Kellogg, Arthur. "Behind the Levees." *Survey*, 58(June 1, 1927): 277–282, 290–292.

Kellogg, Arthur. "Up From the Bottom Lands." *Survey*, 58(July 1, 1927): 360–366.

Kellogg, Paul U. "Drought and the Red Cross." *Survey*, 65(February 15, 1931): 535–538, 572–576.

Kelsey, Carl. *The Negro Farmer*. Chicago, Ill.: Jennings and Pye, 1903.

Kennedy, Renwick C. "The Cotton Kingdom: 1931: I: Six-Cent Cotton, a Southern Tragedy." *New Republic*, 69(December 16, 1931): 129–130.

King, Edward. *The Great South: A Record of Journeys*. Hartford, Conn.: American Publishing Co., 1875.

King, J. Crawford, Jr. "The Closing of the Southern Range: An Exploratory Study." *Journal of Southern History*, 48(February 1982): 53–70.

Kipp, H. A. *Report on the Belzoni Drainage District in Washington County, Mississippi*. U.S. Department of Agriculture, Office of Experiment Stations, Bulletin No. 244. Washington, D.C.: Government Printing Office, 1912.

Kirby, Jack T. *Rural Worlds Lost: The American South, 1920–1960*. Baton Rouge, La.: Louisiana State University Press, 1987.

Kirby, W. S. *The Drummer Boy of the Ozarks; or, Sketches in the Life of Ben Elder*. West Plains, Mo.: Gazette Steam Printer, 1893.

Knox, Thomas W. *Camp-Fire and Cotton-Field: Southern Adventure in Time of War*. New York, N.Y.: Blelock and Co., 1865.

Koch, Lucien. "The War in Arkansas." *New Republic*, 82(March 27, 1935): 182–184.

Kyle, C. H. *Corn Culture in the Southeastern States*. U.S. Department of Agriculture, Farmers' Bulletin No. 729. Washington, D.C.: Government Printing Office, 1916.

Lambert, Roger. "Hoover and the Red Cross in the Arkansas Drought of 1930." *Arkansas Historical Quarterly*, 29(Spring 1970): 3–19.

Lancaster, Bob. *The Jungles of Arkansas: A Personal History of the Wonder State*. Fayetteville, Ark.: University of Arkansas Press, 1989.

Langsford, E. L., and B. H. Thibodeaux. *Plantation Organization and Operation in the Yazoo–Mississippi Delta Area.* U.S. Department of Agriculture, Technical Bulletin No. 682. Washington, D.C.: Government Printing Office, 1939.

Lemann, Nicholas. *The Promised Land: The Great Black Migration and How It Changed America.* New York, N.Y.: Alfred A. Knopf, 1991.

Leslie, James W. *Land of Cypress and Pine.* Little Rock, Ark.: Rose Publishing Co., 1976.

Lesseig, Corey T. "Out of the Mud: The Good Roads Crusade and Social Change in Twentieth Century Mississippi." *Journal of Mississippi History,* 60(Spring 1998): 51–72.

Letter of Gov. Jas. L. Alcorn, and Resolution Memorializing Congress Relative to Overflowed Lands on the Mississippi River. Jackson, Miss.: Kimball, Raymond and Co., 1870.

"The Levee System of Louisiana." *DeBow's Review,* o.s. 8(February 1850): 101–105.

"The Levees of the Mississippi." *DeBow's Review,* a.w.s. 5(February 1868): 162–167.

Long, Lewis E. *Farm Power in the Yazoo–Mississippi Delta.* Mississippi Agricultural Experiment Station, Bulletin No. 295. Starkville, Miss.: Mississippi A & M College, 1931.

Long, Lewis E., and R. S. Kifer. *Farm Practices in South Central Mississippi with Suggested Changes.* Mississippi Agricultural Experiment Station, Bulletin No. 276. Starkville, Miss.: Mississippi A & M College, 1929.

Louisiana. Washington, D.C.: U.S. Railroad Administration, 1919.

"Louisiana—Her Public Lands and Levees." *DeBow's Review,* 10(May 1851): 529–535.

"The Louisiana Levees." *DeBow's Review,* a.w.s. 3(April–May 1867): 469–473.

McBride, Mary G., and Ann M. McLaurin. "The Origin of the Mississippi River Commission." *Louisiana History,* 36(Fall 1995): 398–411.

McCombs, W. A. "The Present Status of Navigation on the Lower Mississippi River." *Journal of Geography,* 24(January 1925): 11–19.

McCrory, S. H. "The Drainage Movement in the United States." In U.S. Department of Agriculture, *Yearbook of the U.S. Department of Agriculture: 1918,* 137–144. Washington, D.C.: Government Printing Office, 1919.

MacDonald, Purificacion, O., and others. "Documentation, Chronology, and Future Projections of Bottomland Hardwood Habitat Loss in the Lower Mississippi Alluvial Plain." Washington, D.C.: Division of Ecological Services, U.S. Department of the Interior, 1979.

McDowall, Sue Ellen P. *Cotton and Jasmine: A Southern Mosaic.* New York, N.Y.: Vantage Press, 1956.

McEathorn, W. J., and others. *A Report Upon the Drainage of the Agricultural Lands of Bolivar County, Mississippi.* U.S. Department of Agriculture, Office of Experiment Stations, Circular No. 81. Washington, D.C.: Government Printing Office, 1909.

McIver, H. M. "Reminiscences of an Arkansas Pioneer as Recorded in 1890." *Arkansas Historical Quarterly,* 17(Spring 1958): 56–67.

McMillen, Neil R. *Dark Journey: Black Mississippians in the Age of Jim Crow.* Urbana, Ill.: University of Illinois Press, 1989.

McNair, A. D. *Crop Systems for Arkansas.* U.S. Department of Agriculture, Farmers' Bulletin No. 1000. Washington, D.C.: Government Printing Office, 1918.

McNair, A. D. *Labor Requirements of Arkansas Crops.* U.S. Department of Agriculture, Department Bulletin No. 1181. Washington, D.C.: Government Printing Office, 1924.

McNair, A. D., and W. B. Mercier. *Lespedeza or Japan Clover.* U.S. Department of Agriculture, Farmers' Bulletin No. 441. Washington, D.C.: Government Printing Office, 1911.

McNeely, J. G., and Glen T. Barton. *Land Tenure in Arkansas: II: Change in Labor Organization on Cotton Farms.* Arkansas Agricultural Experiment Station, Bulletin No. 397. Fayetteville, Ark.: University of Arkansas, 1940.

McNew, Ben B. "Banking, 1890–1970." In *A History of Mississippi* Ed. by Richard A. McLemore, Vol. 2: 312–333. Jackson, Miss.: University Press of Mississippi, 1973.

Marsden, Roger D., and R. P. Teele. *Economic Status of Drainage Districts in the South in 1926.* U.S. Department of Agriculture, Technical Bulletin No. 194. Washington, D.C.: Government Printing Office, 1930.

Mathews, John L. *Remaking the Mississippi.* Boston, Mass.: Houghton Mifflin Co., 1909.

Mayer, August T. *The Cattle Tick in Its Relation to Southern Agriculture.* U.S. Department of Agriculture, Farmers' Bulletin No. 261. Washington, D.C.: Government Printing Office, 1906.

Meade, Melinda S. "The Rise and Demise of Malaria." *Southeastern Geographer,* 20(November 1980): 77–99.

Mims, Sam. "Louisiana's Administration of Swamp Land Funds." *Louisiana Historical Quarterly,* 28(January 1945): 277–325.

"Mississippi." *DeBow's Review,* o.s. 7(July 1849): 38–44.

Mississippi. Washington, D.C.: U.S. Railroad Administration, 1919.

Mississippi Digest: 1818 to Date. St. Paul, Minn.: West Publishing Co., 1979.

"Mississippi Farm Empire: Ex-Field-Hand Johnie[sic] Brown Owns Six-Mile Long Plantation Along Mississippi River Levee." *Ebony,* 6(February 1951): 30–36.

"The Mississippi Swamp." *DeBow's Review,* o.s. 7(July 1849): 53–56.

The Mississippi Valley Flood Disaster of 1927: Official Report of the Relief Operations. Washington, D.C.: American National Red Cross, [1929].

Missouri. Washington, D.C.: U.S. Railroad Administration, 1919.

Missouri: Imperial Mistress of States. Columbia, Mo.: Missouri State Board of Agriculture, 1898.

Missouri Digest: 1821 to Date. St. Paul, Minn.: West Publishing Co., 1952.

Missouri of To-Day. St. Louis, Mo.: Confederate Home Association of Missouri, 1893.

Mitchell, H.[arry] L. *Mean Things Happening in This Land: The Life and Times of H. L. Mitchell, Co-Founder of the Southern Tenant Farmers' Union.* Montclair, N.J.: Allanheld, Osmun and Co., 1979.

Mitchell, H. L., and J. R. Butler. "The Cropper Learns His Fate." *Nation,* 141(September 18, 1935): 328–329.

Mitchell, H. L., and Howard Kester. "Correspondence: Share-Cropper Misery and Hope." *Nation,* 142(February 12, 1936): 185.

Mitchell, Jonathan. "Cabins in the Cotton." *New Republic*, 92(September 22, 1937): 175–177.

Mohler, John R. *Texas or Tick Fever and Its Prevention.* U.S. Department of Agriculture, Farmers' Bulletin No. 258. Washington, D.C.: Government Printing Office, 1906.

Mohler, John R. *Tick Fever.* U.S. Department of Agriculture, Farmers' Bulletin No. 1625. Washington, D.C.: Government Printing Office, 1930.

Moneyhon, Carl H. "Economic Democracy in Antebellum Arkansas, Phillips County, 1850–1860." *Arkansas Historical Quarterly*, 40(Summer 1981): 154–172.

Moneyhon, Carl H. "The Impact of the Civil War in Arkansas: The Mississippi River Plantation Counties." *Arkansas Historical Quarterly*, 51(Summer 1992): 105–118.

Moneyhon, Carl H. "Delta Towns." In *The Arkansas Delta: Land of Paradox* Ed. by Jeannie Whayne and Willard Gatewood, 208–235. Fayetteville, Ark.: University of Arkansas Press, 1993.

Moneyhon, Carl H. *The Impact of the Civil War and Reconstruction in Arkansas: Persistence in the Midst of Ruin.* Baton Rouge, La.: Louisiana State University Press, 1994.

Montgomery, Frank A. *Reminiscences of a Mississippian in Peace and War.* Cincinnati, Ohio: Robert Clarke Co., 1901.

Montgomery, Hon. Isaiah T. "In the Yazoo–Mississippi Delta." *Southern Workman*, 29(November 1900): 644–648.

Moore, John H. *The Emergence of the Cotton Kingdom in the Old Southwest: Mississippi, 1770–1860.* Baton Rouge, La.: Louisiana State University Press, 1988.

Morgan, A. E. *A Preliminary Report on the St. Francis Valley Drainage Project in Northeastern Arkansas.* U.S. Department of Agriculture, Office of Experiment Stations, Circular No. 86. Washington, D.C.: Government Printing Office, 1909.

Morgan, A. E., and others. *A Preliminary Report on the Drainage of the Fifth Louisiana Levee District.* U.S. Department of Agriculture, Office of Experiment Stations, Circular No. 104. Washington, D.C.: Government Printing Office, 1911.

Morgan, A.[lbert] T. *Yazoo; or, on the Picket Line of Freedom in the South.* Washington, D.C.: Press of Rufus Darby, 1884.

Morris, Christopher. *Becoming Southern: The Evolution of a Way of Life, Warren County and Vicksburg, Mississippi, 1770–1860.* New York, N.Y.: Oxford University Press, 1995.

Morse, W. J. "The Soy-Bean Industry in the United States." In U.S. Department of Agriculture, *Yearbook of the U.S. Department of Agriculture: 1917*, 101–112. Washington, D.C.: Government Printing Office, 1918.

Morse, W. J. *The Soy-Bean: Its Culture and Uses.* U.S. Department of Agriculture, Farmers' Bulletin No. 973. Washington, D.C.: Government Printing Office, 1918.

Moulton, Forest R., ed. *Historical Introduction to the Symposium on Malaria.* American Association for the Advancement of Science, Publication No. 15. Washington, D.C.: American Association for the Advancement of Science, 1941.

Munn, Alvin A. "Production and Utilization of the Soybean in the United States." *Economic Geography*, 26(July 1950): 223–234.

Murfree, W. L. "The Levees of the Mississippi." *Scribner's Monthly*, 22(July 1881): 420–432.

Murphy, Helen. "Overflow Notes." *Atlantic Monthly*, 140(August 1927): 223–230.

Murray, Gail S. "Forty Years Ago: The Great Depression Comes to Arkansas." *Arkansas Historical Quarterly*, 29(Winter 1970): 291–312.

Musoke, Moses S. "Mechanizing Cotton Production in the American South: The Tractor, 1915–1960." *Explorations in Economic History*, 18(October 1981): 347–375.

Naison, Mark D. "Black Agrarian Radicalism in the Great Depression: The Threads of a Lost Tradition." *Journal of Ethnic Studies*, 4(Fall 1973): 47–65.

Nash, Gerald D. "Herbert Hoover and the Origins of the Reconstruction Finance Corporation." *Mississippi Valley Historical Review*, 46(December 1959): 455–468.

Nathans, Sydney. "Gotta Mind to Move, A Mind to Settle Down: Afro-Americans and the Plantation Frontier." In *A Master's Due: Essays in Honor of David Herbert Donald* Ed. by William J. Cooper, Jr., and others, 204–222. Baton Rouge, La.: Louisiana State University Press, 1985.

Nelson, Lawrence J. "Oscar Johnston, the New Deal, and the Cotton Subsidy Payments Controversy, 1936–1937." *Journal of Southern History*, 40(August 1974): 399–416.

Nelson, Lawrence J. "Welfare Capitalism on a Mississippi Plantation in the Great Depression." *Journal of Southern History*, 50(May 1984): 225–250.

Newell, F. H. "Mississippi River Floods." *American Review of Reviews*, 75(June 1927): 593–600.

Niebuhr, Reinhold. "Meditations from Mississippi." *Christian Century*, 54(February 10, 1937): 183–184.

Nordhoff, Charles. *The Cotton States in the Spring and Summer of 1875.* New York, N.Y.: D. Appleton and Co., 1876.

Nourse, B. F. "Report on Cotton." *DeBow's Review*, a.w.s. 7(September 1869): 783–795.

"Now the 'Farm Breadline.' " *Literary Digest*, 108(January 24, 1931): 9–10.

Oden, Jack P. "A Century of Agriculture in the Carroll Parishes of Louisiana, 1800–1900." M.A. thesis, Northeast Louisiana State College, Monroe, 1967.

Odom, E. Dale. "The Vicksburg, Shreveport and Texas: The Fortunes of a Scalawag Railroad." *Southwestern Social Science Quarterly*, 44(December 1963): 277–285.

Ogilvie, Leon P. "The Development of the Southeast Missouri Lowlands." Ph.D. dissertation, University of Missouri, Columbia, 1967.

Ogilvie, Leon P. "Governmental Efforts at Reclamation in the Southeast Missouri Lowlands." *Missouri Historical Review*, 64(January 1970): 150–176.

Okey, Charles W. *The Wet Lands of Southern Louisiana and Their Drainage.* U.S. Department of Agriculture, Department Bulletin No. 652. Washington, D.C.: Government Printing Office, 1918.

Oldham, W. S. "Last Days of the Confederacy." *DeBow's Review*, a.w.s. 8(August 1870): 640–653.

Olmsted, Frederick L. *A Journey in the Back Country.* New York, N.Y.: Mason Bros., 1860.

Orndluff, Donald R. *The Hereford in America.* Privately printed, 1957.

Osband, Kent. "The Boll Weevil Versus 'King Cotton'." *Journal of Economic History*, 45(September 1985): 627–643.

Otto, John S. *The Southern Frontiers, 1607–1860.* Westport, Conn.: Greenwood Press, 1989.

Otto, John S. *Southern Agriculture During the Civil War-Era, 1860–1880.* Westport, Conn.: Greenwood Press, 1994.

"Our Great Drought Still With Us." *Literary Digest*, 108(March 21, 1931): 23.

"Our Southern Tormentor—The Mosquito." *DeBow's Review*, o.s. 19(August 1855): 171–185.

Owens, Harry P. *Steamboats and the Cotton Economy: River Trade in the Yazoo-Mississippi Delta.* Jackson, Miss.: University Press of Mississippi, 1990.

Palmer, Robert. *Deep Blues.* New York, N.Y.: Viking Press, 1981.

"The Parish of Concordia." *DeBow's Review*, 11(July 1851): 40–62.

"The Parish of Tensas." *DeBow's Review*, 14(May 1853): 431–436.

Paul, George F. "Welfare Work on a Delta Plantation." *Southern Workman*, 54(July 1925): 314–318.

Pennybaker, J. E. "State Management of Public Roads." In U.S. Department of Agriculture, *Yearbook of the U.S. Department of Agriculture: 1914*, 211–226. Washington, D.C.: Government Printing Office, 1915.

Pennybaker, J. E. "Federal Aid to Highways." In U.S. Department of Agriculture, *Yearbook of the U.S. Department of Agriculture: 1917*, 127–138. Washington, D.C.: Government Printing Office, 1918.

Percy, William A. *Lanterns on the Levee: Recollections of a Planter's Son.* New York, N.Y.: Alfred A. Knopf, 1941.

Pereyra, Lillian. "James Lusk Alcorn and a Unified Levee System." *Journal of Mississippi History*, 27(February 1965): 18–41.

Perkins, Van L. *Crisis in Agriculture: The Agricultural Adjustment Administration and the New Deal, 1933.* University of California Publications in History, Vol. 81. Berkeley, Cal.: University of California Press, 1969.

Pickens, William. *Bursting Bonds: The Autobiography of a "New Negro."* Ed. by William L. Andrews. Bloomington, Ind.: Indiana Univerity Press, 1991.

Piper, C. V., and H. T. Nielsen. *Soy Beans.* U.S. Department of Agriculture, Farmers' Bulletin No. 372. Washington, D.C.: Government Printing Office, 1909.

Pittman, Dan W. "The Founding of Dyess Colony." *Arkansas Historical Quarterly*, 29(Winter 1970): 313–326.

"Plantation Hygiene." *Southern Cultivator*, 15(May 1857): 140–142.

Porter, Horace G. "New Farms in the Mississippi Delta." *Louisiana Rural Economist*, 2(April 1940): 5–7.

Powell, Betty F. *History of Mississippi County, Missouri.* Independence, Mo.: BNL Library Service, 1975.

Powell, Lawrence N. *New Masters: Northern Planters During the Civil War and Reconstruction.* New Haven, Conn.: Yale University Press, 1980.

Pratt, Joseph H. "Good Roads Movement in the South." *Annals of the American Academy of Political and Social Science*, 35(January–June 1910): 105–113.

Preston, Howard L. *Dirt Roads to Dixie: Accessibility and Modernization in the South, 1885–1935*, Knoxville, Tenn.: University of Tennessee Press, 1991.

Prichard, Walter, ed. "A Tourist's Description in 1860." *Louisiana Historical Quarterly*, 21(October 1938): 1110–1214.

Primack, Martin L. "Land Clearing Under Nineteenth Century Techniques." *Journal of Economic History*, 22(December 1962): 484–497.

Primack, Martin L. "Farm Fencing in the Nineteenth Century." *Journal of Economic History*, 29(June 1969): 287–291.

Prunty, Merle, Jr. "The Renaissance of the Southern Plantation." *Geographical Review*, 45(October 1955): 459–491.

Ralph, Julian. "The Old Way to Dixie." *Harper's New Monthly Magazine*, 86(January 1893): 165–184.

"Reclamation of the River Wilds of the Southwest." *DeBow's Review*, 26(March 1859): 239–250.

Redding, R. J. *Essential Steps in Securing an Early Crop of Cotton*. U.S. Department of Agriculture, Farmers' Bulletin No. 217. Washington, D.C.: Government Printing Office, 1905.

Reid, T. Roy. "Public Assistance to Low-Income Farmers of the South." *Journal of Farm Economics* 21(February 1939): 188–194.

Reid, Whitelaw. *After the War: A Southern Tour—May 1, 1865, to May 1, 1866.* Cincinnati, Ohio: Moore, Wilstach and Baldwin, 1866.

Reinders, Robert C. "Dr. James G. Carson's Canebrake: A View of an Ante-Bellum Louisiana Plantation." *Louisiana Historical Quarterly*, 33(October 1950): 353–363.

Relief Work in the Drought of 1930–31. Washington, D.C.: American National Red Cross, 1931.

Reynolds, S. P. *History of Levees and Drainage in Southeast Missouri*. Missouri Agricultural Experiment Station, Bulletin No. 323. Columbia, Mo.: University of Missouri, 1933.

Ridpath, Ben M. "The Case of the Missouri Sharecroppers." *Christian Century*, 56(February 1, 1939): 146–148.

Roark, James L. *Masters Without Slaves: Southern Planters in the Civil War and Reconstruction*. New York, N.Y.: W. W. Norton and Co., 1977.

Robbins, A. "Hunger—1931." *Nation*, 132(February 11, 1931): 151–152.

Roberts, Bobby. "Desolation Itself.: The Impact of the Civil War." In *The Arkansas Delta: Land of Paradox*. Ed. by Jeannie Whayne and Willard Gatewood, 70–93. Fayetteville, Ark.: University of Arkansas Press, 1993.

Rosencrantz, Florence L. "The Rice Industry in Arkansas." *Arkansas Historical Quarterly*, 5(Summer 1946): 123–137.

"The Row Over Feeding Our Hungry Farmers." *Literary Digest*, 108(January 31, 1931): 8–9.

Rowland, Dunbar. *History of Mississippi*. Chicago Ill.: S. J. Clarke Publishing Co., 1925.

Royce, Edward. *The Origins of Southern Sharecropping*. Philadelphia, Penn.: Temple University Press, 1993.

Rubin, Julius. "The Limits of Agricultural Progress in the Nineteenth Century South." *Agricultural History*, 49(April 1975): 362–373.

Rudd, Dan A., and Theo. Bond. *From Slavery to Wealth: The Life of Scott Bond.* Madison, Ark.: Journal Printing Co., 1917.

Russell, E. Z. *Breeds of Swine.* U.S. Department of Agriculture, Farmers' Bulletin No. 1263. Washington, D.C.: Government Printing Office, 1922.

Salmon, D. E. *Hog Cholera and Swine Plague.* U.S. Department of Agriculture, Farmers' Bulletin No. 24. Washington, D. C.: Government Printing Office, 1894.

Sampson, Ernest E. "Half a Century on Grand Prairie." *Arkansas Historical Quarterly*, 14(Spring 1955): 32–37.

Schafer, A. L. "When Hunger Followed Drought." *Survey*, 65(March 1, 1931): 580–583, 627–628.

"The 'Self-Help' Drought-Relief Program." *Literary Digest*, 106(August 30, 1930): 9.

Semple, Arthur T. *Beef Production in the Cotton Belt.* U.S. Department of Agriculture, Farmers' Bulletin No. 1379. Washington, D.C.: Government Printing Office, 1923.

Shaler, N. S. "The Floods of the Mississippi Valley." *Atlantic Monthly*, 61(May 1883): 653–660.

Shugg, Roger W. "Survival of the Plantation System in Louisiana." *Journal of Southern History*, 3(August 1937): 311–325.

Sillers, Florence W., comp. *History of Bolivar County, Mississippi.* Ed. by Wirt A. Williams. Jackson Miss.: Hederman Bros., 1948.

Sillers, Walter, Sr. "Flood Control in Bolivar County, 1838–1924." *Journal of Mississippi History*, 9(January 1947): 3–20.

Simonson, S. E. "Origin of Drainage Projects in Mississippi County." *Arkansas Historical Quarterly*, 5(Fall 1946): 263–273.

Simonson, S. E. "The St. Francis Levee and High Waters on the Mississippi River." *Arkansas Historical Quarterly* 6(Winter 1947): 419–429.

Smith, A. G. *Soy Beans in Systems of Farming in the Cotton Belt.* U.S. Department of Agriculture, Farmers' Bulletin No. 931. Washington, D.C.: Government Printing Office, 1918.

Smith, Jared G. *Alfalfa or Lucern.* U.S. Department of Agriculture, Farmers' Bulletin No. 31. Washington, D.C.: Government Printing Office, 1899.

Smith, J. Russell. *North America.* New York, N.Y.: Harcourt, Brace and Co., 1925.

Smith, J. Sheppard. "Reclamation of Swamp Lands and the Modern Drainage Bond." *Annals of the American Academy of Political and Social Science*, 88(March 1920): 102–113.

Smith, Raymond C. "Ol' Man River's New Homes." *Land Policy Review*, 4(February 1941): 3–8.

Smith, T. Lynn. "The Redistribution of the Negro Population of the United States, 1910–1960." *Journal of Negro History*, 51(July 1966): 155–173.

Smith, T. Lynn, and S. Earl Grigsby. *The Situation and Prospects of the Population in the Black River Settlement, Louisiana.* Louisiana Agricultural Experiment Station, Bulletin No. 319. Baton Rouge, La.: Louisiana State University, 1940.

Snow, Thad. *From Missouri.* Boston Mass.: Houghton Mifflin Co., 1954.

Snyder, Howard. "Negro Migration and the Cotton Crop." *North American Review*, 219(January 1924): 21–29.

Soil Survey Staff. *Soil Taxonomy.* U.S. Department of Agriculture, Agriculture Handbook No. 436. Washington, D.C. Government Printing Office, 1975.

"Some Facts That the Statistics Show." *Implement and Tractor,* 54(July 8, 1939): 26–53, 69, 75–76.

Somers, Robert. *The Southern States Since the War, 1870–1.* New York, N.Y.: MacMillan and Co., 1871.

"The Southern El-Dorado." *DeBow's Review,* a.w.s. 1(March 1866): 253–255.

"Southern Estates in the Market." *DeBow's Review,* a.w.s. 2(December 1866): 667–668.

"The Southwest," *DeBow's Review,* o.s. 19(November 1855): 613–614.

Speakman, Harold. *Mostly Mississippi.* New York, N.Y.: Dodd, Mead and Co., 1927.

Stanley, Dorothy, ed. *The Autobiography of Sir Henry Morton Stanley.* Boston, Mass.: Houghton Mifflin Co., 1937.

Staples, Thomas S. *Reconstruction in Arkansas, 1862–1874.* Columbia University, Studies in History, Economics, and Public Law, No. 245. New York N.Y.: Columbia University, 1923.

"Starvation in Arkansas." *New Republic,* 86(April 1, 1936): 209–210.

Stepenoff, Bonnie. "The Last Tree Cut Down: The End of the Bootheel Frontier, 1880–1940." *Missouri Historical Review,* 90(October 1995): 61–78.

Stephens, Oran. "Revolt in the Delta: What Happened to the Sharecroppers' Union." *Harper's Magazine,* 183(November 1941): 656–664.

Stephens, P. H. "Mechanization of Cotton Farms." *Journal of Farm Economics,* 13(January 1931): 27–36.

Sternitzke, Herbert S. "Impact of Changing Land Use on Delta Hardwood Forests." *Journal of Forestry,* 74(January 1976): 25–27.

Stone, Alfred H. "The Negro in the Yazoo–Mississippi Delta." *Publications of the American Economic Association,* 3, No. 1(1902): 235–278.

Stone, Alfred H. "The Negro Farmer in the Mississippi Delta." *Southern Workman,* 32(October 1903): 457–461.

Stone, Alfred H. "A Plantation Experiment." *Quarterly Journal of Economics,* 19(February 1905): 270–287.

Stone, Alfred H. *Studies in the American Race Problem.* New York, N.Y.: Doubleday, Page and Co., 1908.

Stover, John F. *The Railroads of the South, 1865–1900.* Chapel Hill, N.C.: University of North Carolina Press, 1955.

Stover, John F. *American Railroads.* Chicago, Ill.: University of Chicago Press, 1961.

Stover, John F. *History of the Illinois Central Railroad.* New York, N.Y.: MacMillan Co., 1975.

Strait, Earl D. *Clearing Land.* U.S. Department of Agriculture, Farmers' Bulletin No. 974. Washington, D.C.: Government Printing Office, 1918.

Strauss, Frederick, and Louis Bean. *Gross Farm Income and Indices of Farm Production and Prices in the United States, 1869–1937.* U.S. Department of Agriculture, Technical Bulletin No. 703. Washington, D.C.: Government Printing Office, 1940.

Street, James H. *The New Revolution in the Cotton Economy: Mechanization and Its Consequences.* Chapel Hill, N.C.: University of North Carolina Press, 1957.

Street, James H. *Look Away! A Dixie Notebook*. Westport, Conn.: Greenwood Press, 1977.

"Stricken Arkansas." *Outlook and Independent*, 157(January 31, 1931): 85.

Strickland, Arvah E. "The Plight of the People in the Sharecroppers' Demonstration in Southeast Missouri." *Missouri Historical Review*, 81(July 1987): 403–416.

Stubbs, William C. *A Hand-Book of Louisiana*. New Orleans, La.: New Orleans Picayune, 1895.

Studabaker, Hugh D. *What They "Showed Me" in Southeast Missouri*. Privately printed, 1913.

Surface, G. T. "Rice in the United States." *Bulletin of the American Geographical Society*, 43(July 1911): 500–509.

Swank, Roy. *Trail to Marked Tree*. Ed. by Nolan Porterfield. San Antonio, Tex.: Naylor Co., 1968.

Switzer, J. E. "A Trip to the Mississippi–Yazoo Flood District." *Journal of Geography*, 26(November 1927): 293–298.

Taylor, Alva W. "Tenancy Report Scores System." *Christian Century*, 52(December 23, 1936): 1730–1731.

Taylor, Joel G. *Louisiana Reconstructed, 1863–1877*. Baton Rouge, La.: Louisiana State University Press, 1974.

Templeton, Geo. S. *Beef Cattle Production in Mississippi*. Mississippi Agricultural Extension Station, Bulletin No. 268. Starkville, Miss.: Mississippi A & M College, 1929.

"Terror in Arkansas." *Nation*, 140(February 13, 1935): 174–175.

Thanet, Octave. "Plantation Life in Arkansas." *Atlantic Monthly*, 68(July 1891): 32–49.

Thanet, Octave. "Town Life in Arkansas." *Atlantic Monthly*, 68(September 1891): 332–340.

Thomas, Lewis F. "Basic Factors in Flood Frequency in the Lower Mississippi River." *Annals of the American Academy of Political and Social Science*, 135(January 1928): 1–6.

Thompson, W. R. *The Pasture Book*. Privately printed, 1950.

Tolley, H. R., and L. M. Church. *Tractors on Southern Farms*. U.S. Department of Agriculture, Farmers' Bulletin No. 1278. Washington, D.C.: Government Printing Office, 1922.

Tolnay, Stewart E., and E. M. Beck. "Racial Violence and Black Migration in the American South, 1910 to 1930." *American Sociological Review*, 57(February 1992): 103–116.

Tompkins, Frank H. *Riparian Lands of Mississippi River*. New Orleans, La.: F. H. Tompkins, 1901.

Tracy, S. M. *Mississippi As It Is: A Handbook of Facts for Immigrants*. Jackson, Miss.: Messenger Publishing Co., 1895.

Tracy, S. M. *Corn Culture in the South*. U.S. Department of Agriculture, Farmers' Bulletin No. 81. Washington, D.C.: Government Printing Office, 1898.

Tracy, S. M. *Hog Raising in the South*. U.S. Department of Agriculture, Farmers' Bulletin No. 100. Washington, D.C.: Government Printing Office, 1899.

Tracy, S. M. *Forage for the Cotton Belt*. U.S. Department of Agriculture, Farmers' Bulletin No. 1125. Washington, D.C.: Government Printing Office, 1920.

Trowbridge, J.[ohn] T. *A Picture of the Desolated States.* Hartford, Conn.: L. S. Stebbins, 1868.

True, A. C., prep. *The Cotton Plant.* U.S. Department of Agriculture, Office of Experiment Stations, Bulletin No. 33. Washington, D.C.: Government Printing Office, 1896.

Tugwell, Rexford G. "The Resettlement Idea." *Agricultural History,* 33(October 1959): 159–164.

U.S. Bureau of Agricultural Economics, prep. *New Settlement in the Mississippi Delta.* U.S. Department of Agriculture, Miscellaneous Publication No. 442. Washington, D.C.: Government Printing Office, 1941.

U.S. Census Bureau. *Thirteenth Census of the United States, Taken in the Year 1910: Agriculture, 1909–1910.* Washington, D.C.: Government Printing Office, 1913.

U.S. Census Bureau. *Fifteenth Census of the United States: 1930: Agriculture.* Washington, D.C.: Government Printing Office, 1932.

U.S. Census Bureau. *Fifteenth Census of the United States: 1930: Population.* Washington, D.C.: Government Printing Office, 1932.

U.S. Census Office. *Agriculture of the United States in 1860.* Washington, D.C.: Government Printing Office, 1864.

U.S. Census Office. *Population of the United States in 1860.* Washington, D.C.: Government Printing Office, 1864.

U.S. Census Office. *Statistics of the Population of the United States at the Tenth Census* (June 1, 1880). Washington, D.C.: Government Printing Office, 1883.

U.S. Census Office. *Tenth Census (1880): Report on the Productions of Agriculture.* Washington, D.C.: Government Printing Office, 1883.

U.S. Census Office. *Tenth Census (1880): Report on Cotton Production in the United States.* Washington, D.C.: Government Printing Office, 1884.

U.S. Census Office. *Twelfth Census of the United States, Taken in the Year 1900: Population.* Washington, D.C.: Government Printing Office, 1901.

U.S. Census Office. *Twelfth Census of the United States, Taken in the Year 1900: Agriculture.* Washington, D.C.: Government Printing Office, 1902.

U.S. Congress. House of Representatives. *Reclamation and Rural Development in the South.* 69th Congress, 2nd Session, Miscellaneous House Document No. 765. Washington, D.C.: Government Printing Office, 1927.

U.S. Congress. House of Representatives. *St. Francis, Mo. [Missouri] and Ark. [Arkansas].* 71st Congress, 2nd Session, House Document No. 159. Washington, D.C.: Government Printing Office, 1930.

U.S. Congress. House of Representatives. *White River, Missouri and Arkansas.* 73rd Congress, 1st Session, House Document No. 102. Washington, D.C.: Government Printing Office, 1933.

U.S. Congress. House of Representatives. *Yazoo River, Mississippi.* 67th Congress, 4th Session, House Document No. 597. Washington, D.C.: Government Printing Office, 1923.

U.S. Congress. Senate. Committee on Agriculture and Forestry. *Report of the Committee on Agriculture and Forestry on Conditions of Cotton Growers in the United States.* 53rd Congress, 3rd Session, Senate Report No. 986. Washington, D.C.: Government Printing Office, 1907.

U.S. Congress. Senate. Committee on Education and Labor. *Report of the Committee of the Senate Upon the Relations Between Labor and Capital, and Testimony Taken by the Committee.* Washington, D.C.: Government Printing Office, 1885.

Vaiden, M. G., and others. *Making Cotton Cheaper: Can Present Production Costs Be Reduced?* Mississippi Agricultural Experiment Station, Bulletin No. 290. Starkville, Miss.: Mississippi A & M College, 1931.

Vaughan, Burton E. "Arkansas Makes a Brilliant Recovery." *American Review of Reviews*, 83(June 1931): 90–91, 103.

Venkataramani, M. S. "Norman Thomas, Arkansas Sharecroppers, and the Roosevelt Agricultural Policies, 1933–1937." *Mississippi Valley Historical Review*, 47(September 1960): 225–246.

"A Victim" [George C. Benham]. *A Year of Wreck: A True Story.* New York, N.Y.: Harper and Bros., 1880.

"Wage Workers and Sharecroppers on Mississippi Plantations." *Monthly Labor Review*, 49(November 1939): 1104–1106.

Wall, E. G. *Hand-Book of the State of Mississippi.* Jackson, Miss.: State Board of Immigration and Agriculture, 1885.

Ward, W. F. "The Production of Beef in the South." In U.S. Department of Agriculture, *Yearbook of the U.S. Department of Agriculture: 1913*, 259–282. Washington, D.C.: Government Printing Office, 1914.

Ward, W. F. *Cottonseed Meal for Feeding Beef Cattle.* U.S. Department of Agriculture, Farmers' Bulletin No. 655. Washington, D.C.: Government Printing Office, 1915.

Ward, W. F., and Dan T. Gray. *Beef Production in the South.* U.S. Department of Agriculture, Farmers' Bulletin No. 580. Washington, D.C.: Government Printing Office, 1914.

Wayne, Michael. *The Reshaping of Plantation Society: The Natchez District, 1860–1880.* Baton Rouge, La.: Louisiana State University Press, 1983.

Webb, Pamela. "By the Sweat of the Brow: The Back-to-the-Land Movement in Depression Arkansas." *Arkansas Historical Quarterly*, 42(Winter 1983): 332–345.

Weeks, Linton. *Clarksdale and Coahoma County: A History.* Clarksdale, Miss.: Carnegie Public Library, 1982.

Wells, O. V. "The Depression of 1873–79." *Agricultural History*, 11(July 1937): 237–249.

Welsh, Donald H., ed. "Martha J. Woods Visits Missouri in 1857." *Missouri Historical Review*, 55(January 1961): 109–123.

West's Louisiana Statutes Annotated. St. Paul, Minn.: West Publishing Co., 1988.

The West-Side Delta. Memphis, Tenn: Southern Alluvial Land Association, 1920.

Westbrook, Lawrence. "The Program of Rural Rehabilitation of the FERA." *Journal of Farm Economics*, 17(February 1935): 89–100.

Whatley, Warren C. "Labor for the Picking: the New Deal in the South." *Journal of Economic History*, 43(December 1983): 905–930.

Whayne, Jeannie. "Creation of a Plantation System in the Arkansas Delta in the Twentieth Century." *Agricultural History*, 66(Winter 1992): 63–84.

Whayne, Jeannie. *A New Plantation South: Land, Labor, and Federal Favor in Twentieth Century Arkansas.* Charlottesville, Va.: University Press of Virginia, 1996.

Whayne, Jeannie, and Willard Gatewood, eds. *The Arkansas Delta: Land of Paradox.* Fayetteville, Ark.: University of Arkansas Press, 1993.

White, John W. *Combinations of Enterprises on Plantations in the Lower Arkansas River Delta.* Arkansas Agricultural Experiment Station, Bulletin No. 449. Fayetteville, Ark.: University of Arkansas, 1944.

White, Max R., and others. *Rich Land—Poor People.* Farm Security Administration, Research Report No. 1. Indianapolis, Ind.: Farm Security Administration, Region III, 1938.

White, Walter. "The Negro and the Flood." *Nation,* 124(June 22, 1927): 688–689.

Wiley, Clarence A. "The Rust Mechanical Cotton Picker and Probable Land-Use Adjustments." *Journal of Land and Public Utility Economics,* 15(May 1939): 155–166.

Williams, Franklin, Jr. *Clearing New Land.* U.S. Department of Agriculture, Farmers' Bulletin No. 150. Washington, D.C.: Government Printing Office, 1902.

Williams, Robert C. *Fordson, Farmall, and Poppin' Johnny: A History of the Farm Tractor and Its Impact on America.* Urbana, Ill.: University of Illinois Press, 1987.

Williamson, Frederick W. *Origin and Growth of Agricultural Extension in Louisiana, 1860–1948.* Baton Rouge, La.: Louisiana State University, 1951.

Williamson, Frederick W., and George T. Goodman, eds. *Eastern Louisiana.* Monroe, La.: Historical Record Association, 1939.

Willis, John C. "On the New South Frontier: Life in the Yazoo-Mississippi Delta, 1865–1920." Ph.D. dissertation, University of Virginia, Charlottesville, 1991.

Wilson, Charles M. "Famine in Arkansas." *Outlook and Independent,* 157(April 29, 1931): 595–597.

Winters, John D. "The Ouachita-Black." In *The Rivers and Bayous of Louisiana* Ed. by Edwin A. Davis, 21–27. Baton Rouge, La.: Louisiana Education Research Association, 1968.

Winters, John D. "Secession and Civil War in North Louisiana." In *North Louisiana* Ed. by B. H. Gilley, 159–183. Ruston, La.: McGinty Trust Fund Publications, 1984.

"Woman Flogged: Charge in Arkansas Share-Croppers' Strike Causes Flare-Up." *Literary Digest,* 121(June 27, 1936): 29–30.

Wood, Stephen E. "The Development of Arkansas Railroads: Parts I and II." *Arkansas Historical Quarterly,* 7 (Summer 1948): 103–140; 7 (Autumn 1948): 155–193.

Woodruff, Nan E. *As Rare As Rain: Federal Relief in the Great Southern Drought of 1930–31.* Urbana, Ill.: University of Illinois Press, 1985.

Wren, Christopher S. "The Restless Ballad of Johnny Cash." *Look,* 33 (April 29, 1969): 68–75.

Wright, J. O. *Excavating Machinery: Used for Digging Ditches and Building Levees.* U.S. Department of Agriculture, Office of Experiment Stations, Circular No. 74. Washington, D.C.: Government Printing Office, 1907.

Wright, J. O. *Swamp and Overflowed Lands in the United States: Ownership and Reclamation.* U.S. Department of Agriculture, Office of Experiment Stations, Circular No. 76. Washington, D.C.: Government Printing Office, 1907.

Wyllie, Irvin G. "Race and Class Conflict on Missouri's Cotton Frontier." *Journal of Southern History,* 20(May 1958): 183–196.

The Yazoo–Mississippi Delta. Memphis, Tenn.: Southern Alluvial Land Association, 1920.

"The Year in Agriculture." In U.S. Department of Agriculture, *U.S. Department of Agriculture: 1931*, 1–90. Washington, D.C.: Government Printing Office, 1931.

Yohe, H. S. *Organization, Financing, and Administration of Drainage Districts.* U.S. Department of Agriculture, Farmers' Bulletin No. 815. Washington, D.C.: Government Printing Office, 1917.

Index

About the Author

JOHN SOLOMON OTTO is a Research Fellow with the International Center, Washington, D.C. He is the author of *Southern Agriculture During the Civil War Era, 1860–1880* (Greenwood, 1994), *The Southern Frontiers, 1607–1860* (Greenwood, 1989), *Cannon's Point Plantation, 1794–1860* (1984), and numerous essays in American History and Culture.

ISBN 0-313-28963-8
90000>
EAN
9 780313 289637
HARDCOVER BAR CODE